...om
politics

NUCLEAR POWER

For Lab School students!
Laurence Pringle
2004

LAURENCE PRINGLE

NUCLEAR POWER

FROM PHYSICS TO POLITICS

MACMILLAN PUBLISHING CO., INC. New York COLLIER MACMILLAN PUBLISHERS London

Macmillan Publishing Co., Inc.
866 Third Avenue, New York, N.Y. 10022
Collier Macmillan Canada, Ltd.
Printed in the United States of America

10 9 8 7 6 5 4 3

LIBRARY OF CONGRESS CATALOGING IN PUBLICATION DATA
Pringle, Laurence P
Nuclear power: from physics to politics.
(Science for survival series)
Bibliography: p.
Includes index.
SUMMARY: Surveys the history and development of nuclear power and presents both sides of the controversy surrounding its use.
1. Atomic power—Juvenile literature.
2. Atomic power-plants—Juvenile literature.
[1. Atomic power. 2. Atomic energy] I. Title.
TK9148.P73 333.7 78-27180 ISBN 0-02-775390-5

Also by Laurence Pringle

SCIENCE FOR SURVIVAL SERIES
Ecology
Energy: Power for People
One Earth, Many People:
The Challenge of Human Population Growth
The Only Earth We Have
Our Hungry Earth:
The World Food Crisis
Pests and People:
The Search for Sensible Pest Control
Recycling Resources

EXPLORING AN ECOSYSTEM SERIES
City and Suburb
Estuaries: Where Rivers Meet the Sea
From Pond to Prairie
The Gentle Desert
The Hidden World: Life Under a Rock
Into the Woods
This Is a River

Dedicated to my parents,
Laurence Erin Pringle
and Marleah Rosehill Pringle

CONTENTS

FOREWORD 1

ONE NUCLEAR ENERGY, PERHAPS 5

TWO HOW SAFE ARE NUCLEAR REACTORS? 29

THREE DANGERS IN THE FUEL CYCLE 53

FOUR COSTS, BREEDERS AND BOMBS 75

FIVE NUCLEAR POLITICS 101

GLOSSARY 121

FURTHER READING 125

INDEX 131

The Three Mile Island Nuclear Plant, with cooling towers, near Harrisburg, Pennsylvania.

FOREWORD

The prospects of nuclear power once seemed very bright. It was called the energy of the future. Most people believed that it was safe, clean, and cheap. Now these beliefs are challenged. Increasing numbers of people doubt the wisdom of developing nuclear energy much further. Indeed, some people believe that existing nuclear generating plants should be phased out as soon as possible.

The controversy continues. The views of both sides are supported by experts, studies, and statistics. At one time or another, each side has been heard to say of its opponents, "The trouble with those people is that they think their ignorance is knowledge."

In this controversy (as well as in others), it is difficult to find a well-informed person who is also neutral. This book is not neutral either. However, I have tried to present views and arguments from both proponents and opponents of nuclear energy. Those readers who want to dig deeper will find a full range of information and opinion in the books, articles, and other sources listed at the back of this book.

It would take many volumes to describe completely the complex story of nuclear-energy development. This book gives an overview of that story. It attempts to clarify the major issues of the controversy—a matter of great importance not just for the late twentieth century but for many generations beyond.

—*Laurence Pringle*

NUCLEAR POWER

In the late 1970s, nuclear power provided about one-eighth of the electricity used in the United States.

1

NUCLEAR ENERGY, PERHAPS

Just a century ago the United States provided all of its own energy. Our needs were not great, and the major fuel was wood. By the late 1970s, however, our highly mechanized and highly wasteful society consumed the equivalent of seven gallons of oil per person every day. But the Petroleum Age is coming to an end. Oil, natural gas, and gasoline (which is made from oil), the fuels which now provide about three-quarters of the power in industrialized nations, will have to be replaced.

Already in the United States there have been temporary but troublesome shortages of natural gas and gasoline in some regions, and rising energy prices everywhere. The United States has become increasingly dependent on foreign oil, importing nearly half of its supplies in 1978. Outside of granting statehood to oil-rich nations (as one wit suggested), there is no quick way for the United States to become less vulnerable to sudden price rises and possible boycotts imposed by oil-exporting countries.

We need to use energy more wisely and develop new sources. Most experts agree that our future needs will be met by a combination of fuels, mostly by coal, solar, and perhaps nuclear energy.

"Perhaps nuclear energy" is a phrase that offends people on both sides of the controversy over this potential source of electric power. One side believes that the development of nuclear power is vital and inevitable. The other believes that nuclear power is extremely dangerous and the only thing inevitable about it is its demise. As in any controversy, there are also more moderate views on both sides.

Nuclear advocates argue that this form of energy is much cleaner, safer, and cheaper than such fossil fuels as oil and coal. They say that we have no choice—the demand for electricity will more than double by the year 2000. In Europe and other parts of the world which lack much coal, nuclear power seems to be even more of a necessity than in the United States. Moreover, proponents of nuclear energy contend that while the need for electric power can be reduced somewhat by using what we have more wisely, conservation alone will not bridge the gap until alternate sources of energy are developed, and these alternatives, especially energy from the sun, will not be of much help until well into the next century.

Opponents of nuclear power are much more optimistic about the development of solar energy and other alternatives. They also say that we can fill our need for new energy from the fifty percent of our energy budget we currently waste. This "treasure"—found in use of home insulation, more fuel-efficient transportation and manufacturing processes, less throwaway packaging, and many other conservation measures—would greatly reduce the need for more new power plants of any kind. Not only is nuclear power unnecessary, they say, it is also unreliable and unsafe. At the heart of their concern is the safety issue. The risks to human life from every stage in the nuclear fuel cycle are simply too great.

The nuclear-energy controversy is not going to be resolved easily. Both sides have considerable political power. Although they use mostly cold statistics and technical jargon in their arguments, both sides are often motivated by strong emotions. And while reasons differ, everyone agrees that decisions about nuclear-energy development are matters of utmost importance to the future of the United States and the world. To understand this complex issue it is important to understand the history and development of nuclear power.

The Nuclear Age dawned in 1945, when the United States dropped atomic bombs on two Japanese cities, Nagasaki and Hiroshima. The bombing had at least two effects: The Japanese surrendered, bringing World War II to a swift end; and many people in the United States, especially scientists and officials involved in the development of atomic bombs, were awed and frightened and filled with guilt over the death and destruction they had caused.

Guilt is a very unpleasant feeling. People go to great lengths to avoid feeling guilty, or to ease their guilt in

Those responsible for the use of nuclear weapons wanted to show that nuclear power could have positive effects.

some way. Some psychologists and historians believe that this feeling of guilt, more than any other reason, led the United States government to invest billions of dollars to develop peacetime uses of nuclear energy, in a kind of crusade to show that nuclear energy could be a force for good.

In 1946, the Atomic Energy Commission (AEC) was created. The commission had complete control of nuclear materials and development. For a decade, however, the military use of nuclear power was emphasized, because of the arms race with the Soviet Union.

Research for military purposes, particularly the development of nuclear-powered submarines, suggested that atomic energy had commercial possibilities. In 1954, the AEC was empowered by Congress to give information to private industry and to encourage the development of atomic energy for generating electricity. The AEC could also grant licenses for nuclear power plants. Thus this federal agency had two jobs—to promote nuclear energy, and also to regulate it and protect the public from its dangers. There was a basic contradiction between the two tasks; it was like spurring a horse on while holding back on the reins. History shows that the job of promotion won over the job of regulation.

Since atomic energy was born in military secrecy and was also a complex new technology, there was an air of mystery about it. The staff and officials of the AEC seemed to believe that it was in the public's interest for the agency to shield citizens from alarming or disturbing news. What people heard, for the most part, was that atomic energy was safe and clean and in good hands with the AEC watching over it.

This image was tarnished somewhat in the late 1950s, when nuclear weapons were being tested aboveground. Radioactive particles from these tests spread through the atmosphere and fell to earth. People worried about this invisible but detectable fallout. The AEC maintained that fallout was harmless. It kept secret the results of studies which proved otherwise. When this information finally reached the public, the AEC was shown for the first time to be covering up and glossing over possible dangers of nuclear energy. It was not the last time.

From 1956 to 1963, public concern about atomic power was mainly focused on the issue of fallout. A nuclear test ban treaty with the Soviet Union ended most of that worry. With their main concern about nuclear energy eased, most people looked forward to the benefits of the peaceful atom.

These benefits were slow in coming. Electric utilities had been urged by the AEC to "go nuclear" but hesitated because the financial risks seemed too great. There were the usual uncertainties of starting in a new and untried technology. Nuclear energy had another sort of financial risk, having to do with insurance.

Ordinarily a business must pay for any damages caused by its goods or services. Protection against this liability, as it is called, is usually provided by insurance. The utilities found, however, that no insurance company was interested in issuing a policy that would pay for all of the damage that might result from an accident at a nuclear power plant. In 1956 an insurance company executive said, "Even if insurance could be found, there is a serious question whether the amount of damage to persons and property would be worth the possible benefit accruing from atomic development."

Then, as now, many people link the words "atomic" and "nuclear" with the word "bomb." Actually the kinds of nuclear power plants now used in the United States cannot explode like a bomb; but a major accident would still be very costly in both lives and property. In 1956 the AEC authorized the first study of the potential risks of nuclear power. The study was conducted at the AEC's Brookhaven National Laboratory on Long Island, New York, and was published in 1957.

According to the Brookhaven Report, as it was called, a major accident at a rather small nuclear power plant located thirty miles from a city could cause 3,400 deaths, 43,000 serious injuries, and $7 billion in property damage. The report did not include any estimates of deaths from cancer caused by radioactivity which might occur many years later.

The AEC argued that these figures were meaningless because extraordinary safety measures could prevent such an accident from happening. But the electric utilities and the insurance companies continued to hesitate. Damage liability was a major obstacle to the development of commercial nuclear power. The obstacle was removed when a special law was passed by Congress in 1957. The Price-Anderson Act protected utilities from full liability for nuclear disaster. According to this law, victims of a nuclear accident could be paid no more than a total of $560 million—much less than the estimated cost of a major nuclear accident. Moreover, most of this limited liability would be paid by the federal government, not by utilities or manufacturers of nuclear equipment.

With the liability obstacles removed, utilities began to apply for licenses to build and operate nuclear power plants. The AEC smoothed their path. It built a demonstration nuclear power plant in Shippingport, Pennsylvania, which began producing electricity in 1958. The AEC also spent billions of dollars on research and provided financial help for the first several large commercial nuclear power stations. There were other incentives. Manufacturers of nuclear equipment, such as Westinghouse and General Electric, sold some of their early reactors at a loss, confident that the money would be recouped later. Nuclear energy began to seem like a sure bet, a wonderful new technology that could be highly profitable.

Nuclear development gained momentum during the 1960s. Scores of plants were built, dozens more were ordered, and in 1968 the AEC predicted that a thousand nuclear power plants would be operating by the year 2000. To people in the business and to most of the general public, it seemed that a golden Nuclear Age lay just ahead.

A few years later the AEC had been abolished, utilities had canceled orders for more than a hundred nuclear

Government encouragement of nuclear power included construction of this plant beside the Ohio River in Shippingport, Pennsylvania.

plants, and the very future of nuclear power in the United States seemed in doubt. There are several reasons for this extraordinary change, but one is quite simple: Many people had concluded that nuclear energy was potentially too dangerous an enterprise to be left to industry and government "experts."

Some of those concerned were nuclear physicists and engineers whose training enabled them to find flaws in the studies and optimistic statements of the AEC. Others were lawyers, housewives, students, and other citizens. To appreciate why these people have tried to halt or slow down nuclear development, we must understand some basic facts about nuclear power and its special characteristics.

THE WORLD OF NUCLEAR FISSION

The word "atomic" refers to atoms, bits of matter that are the smallest units of an element. Tiny atoms make up our bodies, all living things, rocks, soils, water, air, and all other matter on earth. The word "nuclear" refers to the center or nucleus of an atom. These centers—nuclei—are made up of parts called neutrons and protons. Much smaller parts, called electrons, travel in orbits around a nucleus, like planets circling a star.

Scientists once thought that atoms could not be broken apart. (The word "atom" means indivisible.) Now we know that the nuclei of atoms can be split (fissioned). When this happens in very heavy nuclei, such as uranium, tremendous amounts of heat energy are released. Fission of

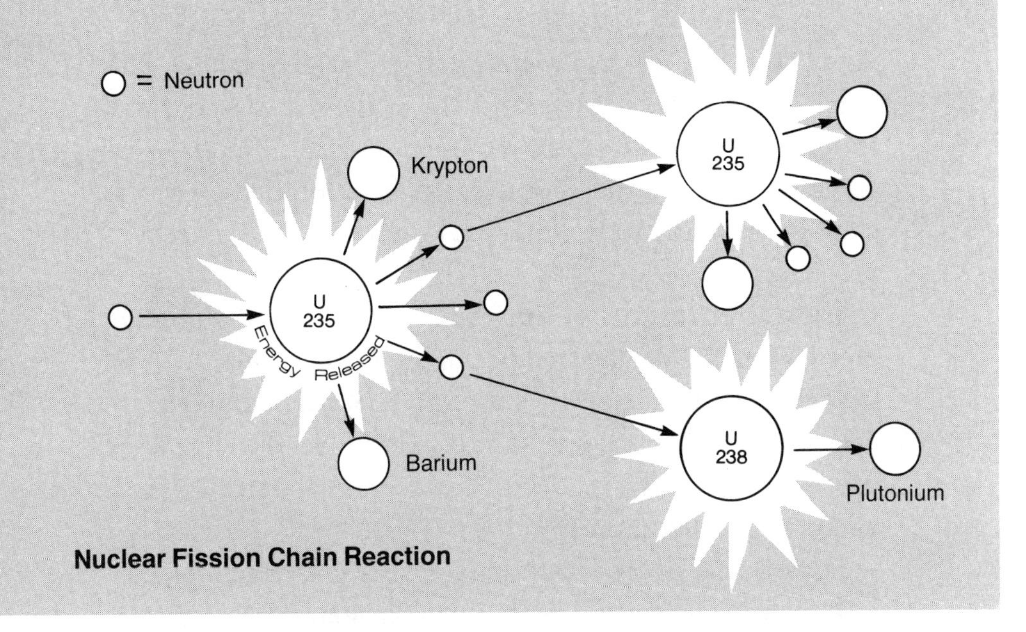

Struck by a neutron, a uranium-235 atom splits and gives off energy and three neutrons, which strike other uranium atoms and cause them to fission, too.

one uranium atom, for example, produces millions of times more energy than the burning of one carbon atom—the energy source in coal.

The destructive force of atomic bombs comes from the fission of atomic nuclei. So does the electricity from a nuclear power plant. Uranium-235 is the basic fuel of today's nuclear plants. When atoms of uranium-235 are bombarded by neutrons, their nuclei split apart and release energy. The uranium atoms are converted into lighter atoms (called fission products) such as strontium and iodine. They also release more neutrons, which collide with other uranium-235 atoms and cause them to fission, and these new fissions produce more neutrons so the process is repeated over and over in a chain reaction. This process can be controlled so that a steady supply of heat energy is produced.

Except for its fuel, a nuclear power plant is similar in many ways to one that uses coal, oil, or natural gas. In all power plants heat energy is used to turn water into steam. The force of the steam striking the blades of a turbine causes a generator to turn. The generator converts heat energy to electrical energy.

Fission takes place in a reactor chamber which is usually more than forty feet high, with steel walls at least eight inches thick. The reactor of a large power plant contains about a hundred tons of uranium. This fuel comes in the form of uranium oxide pellets. Each half-inch pellet can produce energy equal to about twelve barrels of oil or three tons of coal. The pellets are packed inside metal tubes called fuel rods. Bundles of fuel rods are lowered into a reactor and arranged vertically in racks so that water can circulate among them.

The world's most common kind of reactors are called light-water reactors or pressurized-water reactors. The water in these reactors cools the fuel and also slows down the neutrons released by fission. In these types of reactors the neutrons must be slowed for a chain reaction to occur. The number of atoms that fission during any period of time is controlled by rods of boron, cadmium, or another sub-

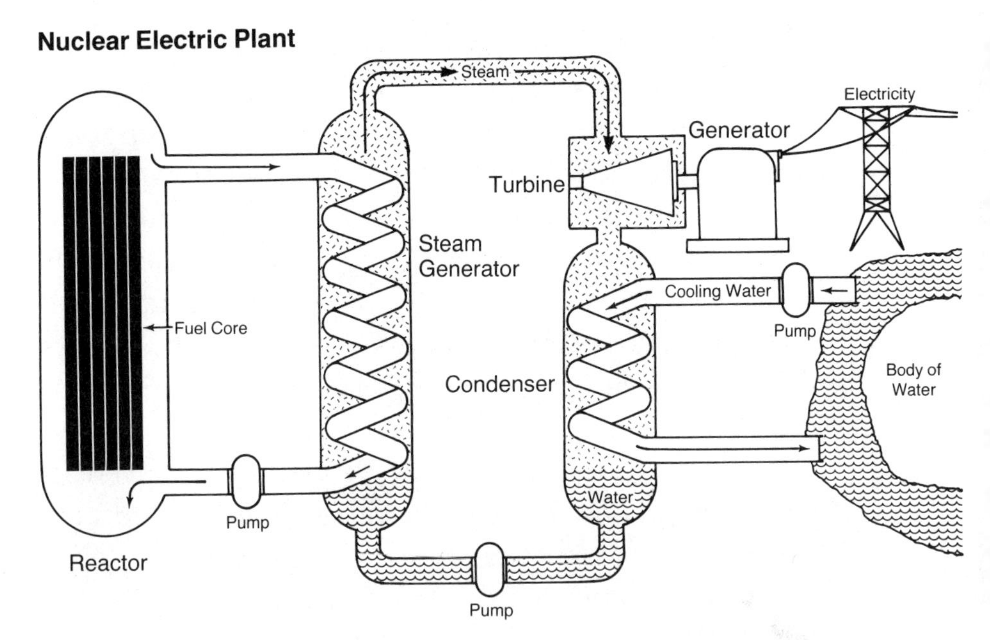

The basic parts of a light-water reactor and electric power plant

stance that absorbs neutrons. By raising or lowering these control rods, operators of a reactor can regulate the amount of heat energy produced. In an emergency, fission can be halted by inserting the control rods among the fuel rods in a reactor.

The water that circulates in a reactor is heated to about 550°F (287°C). Its heat is used directly or indirectly to make steam which then powers a generator. Since the water may contain some radioactivity, it must be confined to the power plant. A separate water supply and circulation system is used to absorb waste heat. This water is usually taken from a river, lake, or ocean, used as coolant, then returned. In a large nuclear plant about 1.2 million gallons of cooling water are used each minute. Powerful pumps and a system of pipes and valves are needed to circulate the cooling water, as well as the water within the reactor.

Nuclear reactors have emergency cooling systems, designed to pump water into the core of the reactor vessel if the normal coolant is lost. Also, the steel and concrete reactor building is another line of defense if the reactor chamber leaks or bursts.

Safety measures like these are stressed in the advertisements of utilities that have nuclear reactors. Without a doubt the nuclear-power industry is subjected to tough safety rules, compared to other industries. Nevertheless, there is still doubt about the safety of reactors. There is also concern about dangers in *all* of the steps in the nuclear fuel cycle, before the fuel reaches a reactor and especially afterward. Nuclear fuel's unique dangers are discussed later in this chapter.

The fuel cycle begins at uranium mines. At mills near the mines uranium ore is crushed, ground, and treated chemically so that the uranium is separated from other materials. The final milling product is called yellowcake. It is taken to uranium-conversion plants where it is chemically treated again and changed to a gas called uranium hexafluoride. Next comes the vital enrichment step, in which the uranium hexafluoride gas is treated so that it contains three percent uranium-235. Natural uranium is of little use in a reactor because it contains less than one

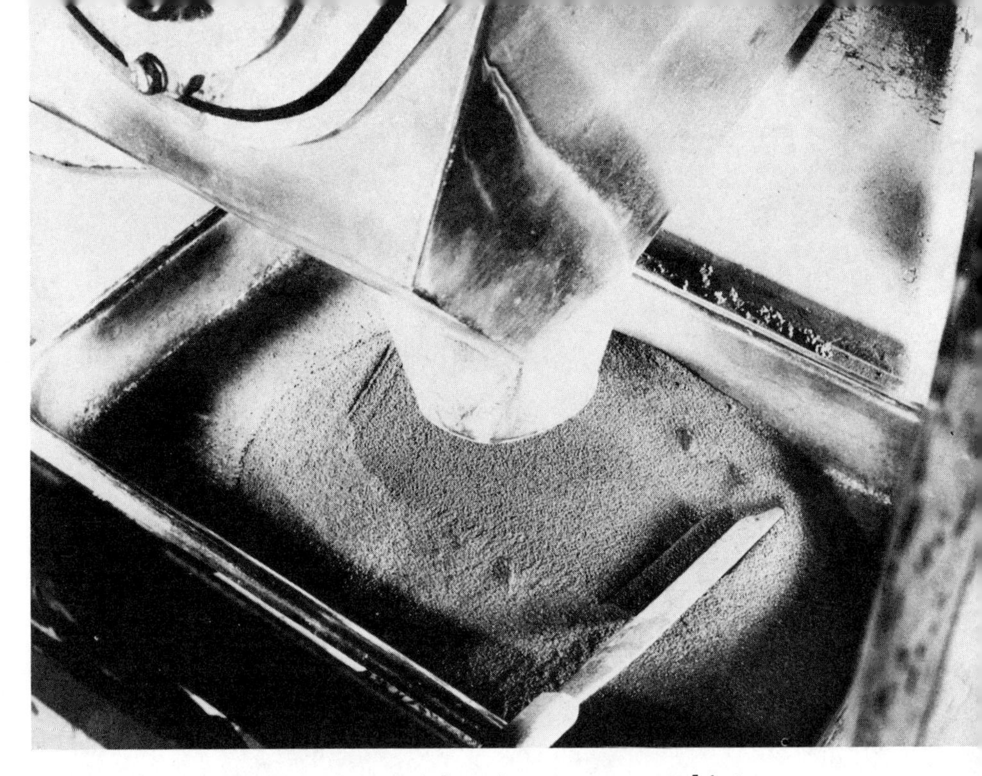

Granules of enriched uranium (above) are compacted into fuel pellets (below), each of which can produce as much electric power as three tons of coal.

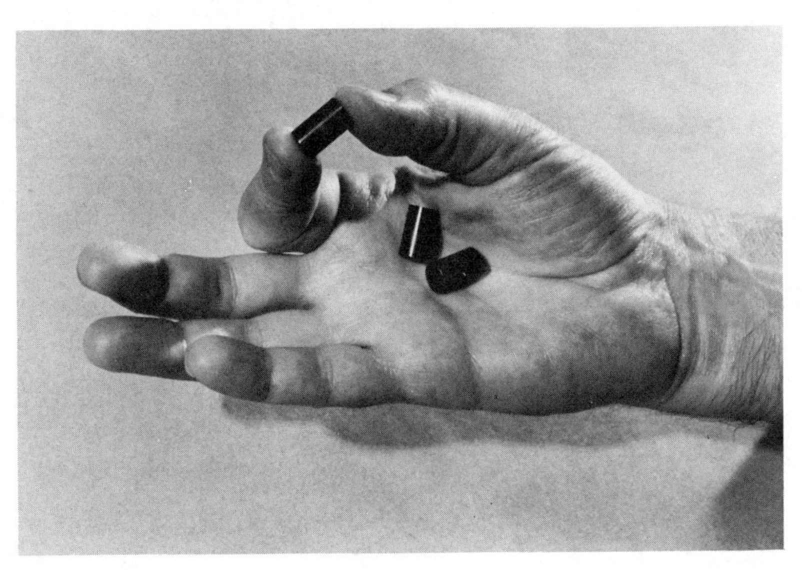

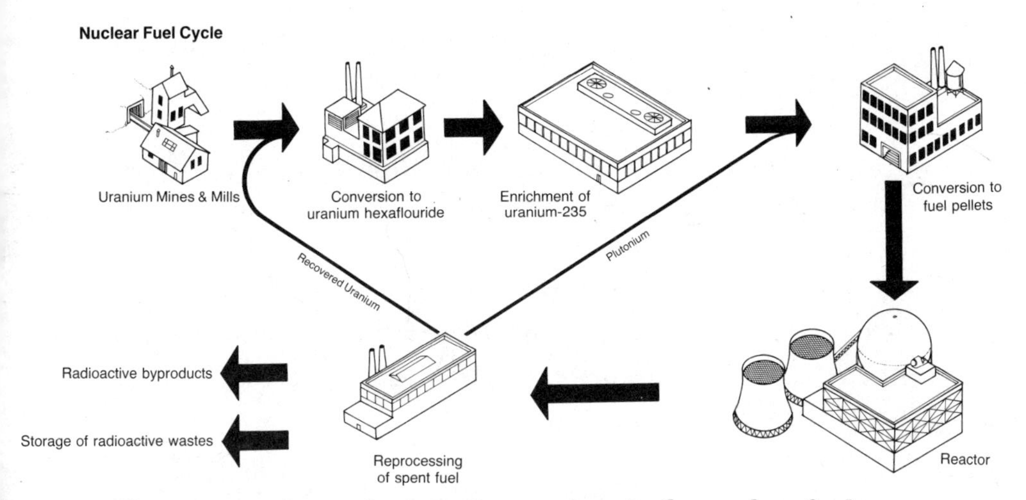

The reprocessing and waste storage steps in the nuclear fuel cycle are beset by unresolved technical and political problems.

percent uranium-235. Enrichment changes it to reactor fuel that can be used in light-water reactors.

Now the fuel is almost ready. The enriched gas is converted to solid pellets of uranium oxide and placed in fuel rods. Bundles of these rods are shipped to reactors. When nearly all of the uranium atoms in a bundle of fuel rods have fissioned, the old fuel is replaced. But it is no ordinary spent fuel. It contains plutonium and other byproducts which are highly poisonous.

Spent fuel has so far been stored near reactors while scientists and engineers try to devise safe long-term

Fuel pellets are placed inside metal rods for use in reactors.

storage. Before final storage, spent fuel may be put through reprocessing—an expensive and dangerous procedure that chemically removes unfissioned uranium and plutonium which can again be used as reactor fuel. But with or without reprocessing, fission byproducts must be stored out of contact with people for many centuries.

Nuclear fuel is obviously unlike such fossil fuels as coal and oil. It is important to recognize its special characteristics because they are at the heart of the opposition to nuclear development.

RADIOACTIVITY

Uranium and a few other elements are naturally radioactive. This means that their atoms spontaneously give off energy. They decay and change to atoms of other, lighter elements.

Radioactive energy is emitted as alpha particles, beta particles, or gamma rays. Alpha particles are the nuclei of helium atoms. They can be stopped by a sheet of paper. Beta particles are electrons. They can be stopped by thin metal. Both kinds of particles, if swallowed or inhaled, can cause damage to internal organs of humans. Gamma rays can pierce thick barriers. (They are used for dental and medical x-rays.) None of this radiation can be seen or felt, but it can still harm living things.

Radioactive elements decay at different rates. The rate is called an element's half-life—the time required for half of the original amount of an element to decay to other

Some Radioactive Substances and Their Half-Lives

element	half-life
Uranium-235	.7 billion years
Plutonium-239	24,000 years
Radium-226	1,620 years
Strontium-90	30 years
Cobalt-60	5.3 years
Iron-55	2.9 years
Iodine-131	8 days
Bismuth-214	19.7 minutes

atoms. Strontium-90, for example, has a half-life of thirty years. After thirty years a pound of strontium-90 would have become a half pound of strontium-90 and a half pound of decay products. After another 30 years only a quarter pound of strontium-90 would remain.

After ten to twenty half-lives pass, only a tiny fraction of the original amount of an element remains. For all practical purposes, at this point the radioactive element can be considered harmless to humans. In the case of strontium-90, this means that as much as six hundred years must pass before a quantity of this element would be virtually harmless.

Scientists use a unit called the rem to match biological damage with a specific dose of radiation. For example, a single dose of 400-500 rem will kill about half of the people exposed to this much radiation. A dose of 100 rem will make most people exposed to it sick.

We know a lot about the effects of single, large doses of radiation. But what about lower doses received over a long period of time? This is one of the more controversial issues in arguments about nuclear energy.

People have always been exposed to low doses of radiation. We take some radioactive atoms into our bodies as we eat, drink, and breathe because there are natural sources of radioactivity on earth. Some comes from uranium and other elements in the earth's crust. Some are the result of cosmic rays from space.

The effects of any low-level radiation—natural or human-made—are hard to assess. One difficulty is the long time required for damaging effects to be recognized. Radiation may cause cancer, but as many as thirty years may pass before the disease is detected in a person's body. Even then there is no way to identify the cause, since both cancer and leukemia (cancer of the blood) can be caused by factors other than radiation.

Radiation damage of other kinds may also be hidden for years. Radiation may alter the genetic material in sex cells. This can affect the survival of a child before birth, and can also cause such birth defects as heart disease and brain damage. But there is no way to tell whether any individual case of genetic damage was caused by radiation.

Nevertheless, animal tests in laboratories demonstrate the dangers of radiation. Opponents of nuclear development are especially concerned about radiation danger in

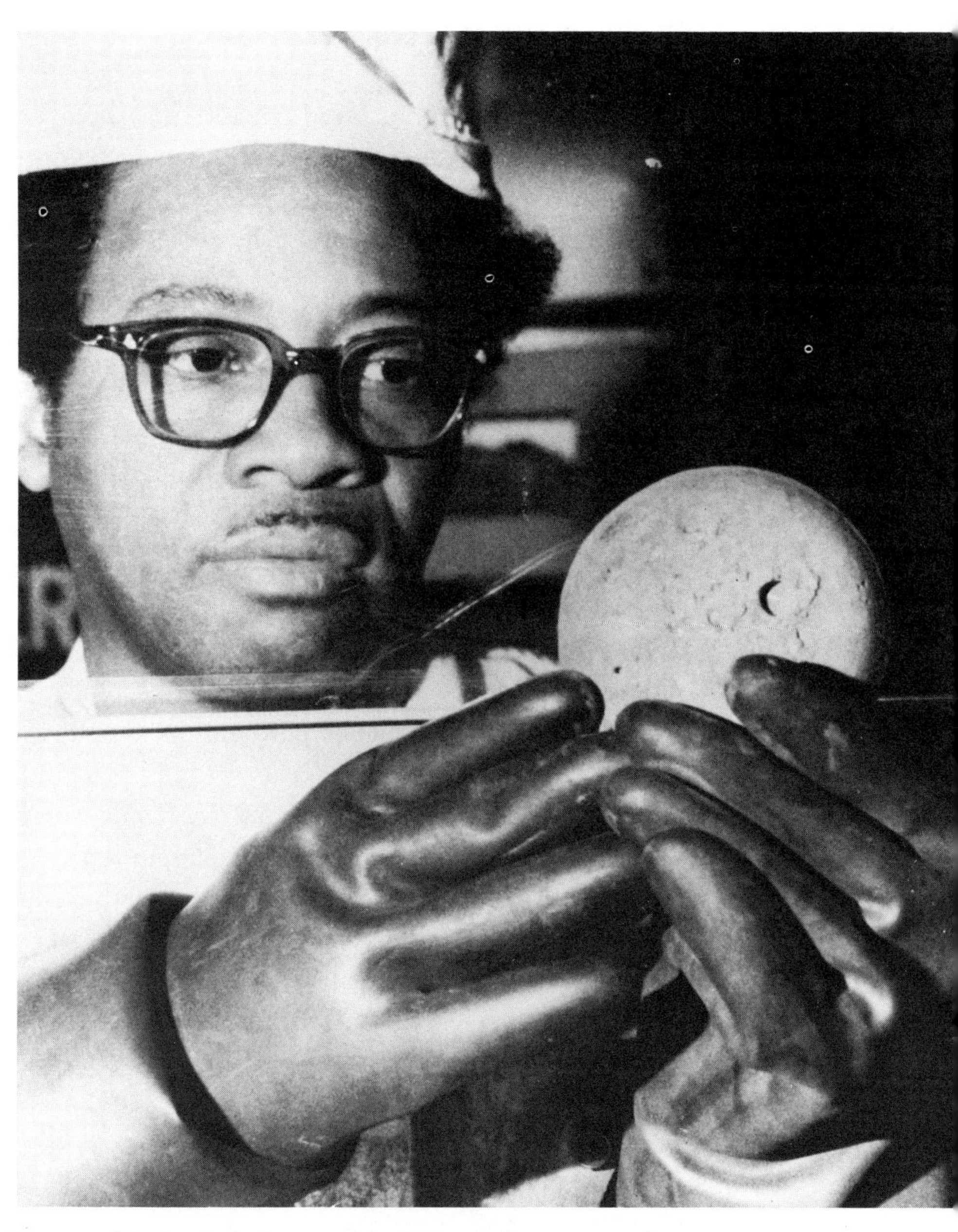

Protected by leaded glass and lead-lined gloves, a nuclear technician examines a piece of plutonium.

the future, when hundreds of large reactors may be operating. Particularly worrisome is the breeder reactor and its fuel, plutonium.

To proponents of nuclear power, the development of breeder reactors is inevitable, since supplies of fissionable uranium are dwindling and a breeder can actually produce more fuel than it uses. The fuel it produces is plutonium-239. Until 1940 there was virtually no plutonium-239 on earth. It is a human-made element, a byproduct of uranium fission.

Plutonium-239 is often called the most dangerous substance on earth. It emits alpha particles. A tiny speck of plutonium-239—as little as three millionths of a gram—can cause lung cancer. One pound of plutonium-239 contains enough "specks" to kill nine billion people. And its half-life of 24,000 years makes plutonium-239 a threat to many future generations.

Like so many other issues in the nuclear controversy, the danger of plutonium is a matter of dispute. Nuclear proponents believe that the danger has been greatly exaggerated. Alpha particles cannot penetrate human skin. Plutonium must be inhaled, swallowed, or enter the body through a cut or other wound in order to be harmful. The image of billions of people dying from a pound of plutonium is horrifying, but this could only happen if the airborne plutonium particles were inhaled by all these people. Nuclear advocates are confident that extraordinary precautions will protect the public.

Arguments about plutonium, and fears of it, may be with us for many years, since the long-term future of nuclear power will depend on breeder reactors and plutonium fuel. The AEC once predicted that hundreds of breeders would be operating in the United States in the twenty-first century. Many people now doubt whether this will ever happen.

There are many complex economic, social, and political factors involved in the nuclear controversy which will play a large part in determining the fate of the "peaceful atom."

Loading bundles of fuel rods into an Iowa reactor

2
HOW SAFE ARE NUCLEAR REACTORS?

"Through safety research and tests, the atomic power industry is continuously strengthening the most important safeguard any industry has—namely, knowledge of the causes and consequences of accidents and of the dependability of safeguards. Learning about accidents before they occur is part of the basic fabric of the safety of atomic power."

These soothing words are quoted from a booklet, *Atomic Power Safety,* distributed to the public by the Atomic Energy Commission in 1964. It is one sample of the public information that was issued by the AEC. In the opinion of some people, the "information" is a lie. At best, it is misleading. As people received a more accurate picture of reactor safety they came to distrust such optimistic views. Today, reactor safety lies at the heart of public concern about nuclear energy.

Nuclear power plants routinely give off some radiation into the air, and in their cooling water. The amounts of such releases have been reduced a lot over the years. In 1954 the National Committee on Radiation Protection stated that human exposure to 36 rem a year would be safe. By 1969, however, as a result of pressure from people outside the government and the nuclear-power industry, the AEC adopted the standard that a person should receive no more than 170 millirem (0.17 rem) of radiation in a year. At that time some scientists claimed that if all persons in the United States were exposed to this amount of radiation the result would be at least "32,000 additional cases of fatal cancer plus leukemia per year, and this would occur every year."

The AEC and the nuclear-power industry disputed these figures. A committee of the National Academy of Sciences studied the matter. Its 1972 report estimated that radiation exposure to 170 millirem would cause 3,000 to 15,000 cancer deaths annually. The committee concluded that an exposure standard of 170 millirem was dangerously high.

In 1975 the Environmental Protection Agency was given the responsibility for setting radiation standards. Early in 1977 it set the annual allowable radiation dose to the public from nuclear power operations at 25 millirem. It is the responsibility of the Nuclear Regulatory Commission (which took over some of the duties of the AEC) to see that this standard is met throughout the entire nuclear-fuel cycle, not just at reactors.

Some proponents of nuclear energy believe that these standards are unnecessarily strict. Their opponents say

Trying to allay public concern about safety, the Atomic Energy Commission distributed this photograph which shows inspection of reactor control rod devices at Browns Ferry, Alabama.

that there may be *no* safe dose of radiation for humans, and that the harm caused by low radiation levels may be underestimated. This controversy is bound to continue until much more is known about the effects of radiation.

The main concern about reactors, however, is not their steady emission of small levels of radiation but the possibility of a major accident that would loose large amounts of radioactive materials into the air. The most feared accident is called a LOCA—loss-of-coolant accident. Pressurized water must keep circulating through a reactor's fuel core to keep it from overheating. One likely cause of a LOCA would be a break in one of the pipes carrying cooling water. Since the water in a reactor is hot (550°F) and pressurized, it would all burst as steam out of a broken pipe.

The reactor could be shut down almost instantly, as control rods were inserted, bringing uranium fission to a halt. But fission products would continue to emit heat as they undergo radioactive decay. If no emergency coolant reached the reactor, the fuel core would heat rapidly to a temperature of 4,000°F (2204°C) or higher. Within the span of three to twenty minutes, the reactor core would become a molten blob that would begin to burn its way through steel and ten or more feet of concrete to the earth beneath the power plant. This was named the China syndrome, in jest, as engineers imagined the hot blob burning through the center of the earth. Realistically, the molten fuel core would probably react with water in the soil.

Reactors are not bombs, but they have the potential of

causing great damage. A typical large reactor may contain as much radioactive material as a thousand bombs of the kind that destroyed Hiroshima. As nuclear fuel fissions, the numbers of uranium-235 nuclei decrease. Some absorb neutrons and become plutonium-239, a fission byproduct which may also change to other forms of plutonium. A variety of other radioactive elements also form. Some are gases, such as krypton and xenon. The half-lives of the elements in this radioactive inventory, as it is called, vary from a few seconds to many thousands of years.

There has been a tremendous amount of speculation and study about a loss-of-coolant accident. Engineers and scientists have tried to anticipate all of the difficulties, and to design safety systems to overcome them. And, beginning with the 1957 Brookhaven Report, there have been three comprehensive studies of the possible consequences of a major reactor accident.

In 1964 employees at the AEC's Brookhaven National Laboratory conducted the second reactor accident study. The results were not encouraging. Bigger reactors were being built. The estimates of potential damage from a major accident rose accordingly: 45,000 deaths, 100,000 injuries, and property damage of $17 billion or more spread over an area the size of Pennsylvania. The AEC scientists who prepared the update study suggested that the liability level of the Price-Anderson Act could be increased forty times.

However, this opinion and the grim figures of the study

itself did not reach the public for more than seven years. The Steering Committee of the AEC met with representatives of the Atomic Industrial Forum, an industry group which promotes nuclear power. The Atomic Industrial Forum urged that the results of the accident study be kept secret, and it was.

Rumors about the suppressed nuclear-accident report lead to inquiries. In reply to requests from the public and members of Congress, the AEC's response was "no new report is in existence or contemplated." Nine years later, in 1973, faced with law suits under the terms of the new Freedom of Information Act, the AEC released more than two thousand pages of the "nonexistent" report, including memos and letters which reflected the agency's desire to keep the report secret.

People who doubted the wisdom of nuclear development were not reassured by these revelations. The release of other AEC documents also revealed that the agency seemed more concerned about the public relations impact of safety studies than the actual safety of reactors.

Safety researchers employed by the AEC were also dissatisfied and worried. A 1970 AEC document listed 139 unresolved safety questions. Forty-four of them were labeled "very urgent, key problem areas." Nevertheless, vital safety research was postponed, and some funds marked for safety studies on water-cooled reactors were diverted to development of the breeder reactor.

By the early 1970s some AEC employees were privately expressing grave doubts about reactor safety. They chose to remain anonymous for fear of losing their jobs. An engineer at the National Engineering Laboratory in Idaho said, "This is being advertised as a no-risk business and that's not true. . . . Maybe it's time the AEC told the public that if people want to turn the lights on they are going to expect to lose a reactor now and then, and possibly suffer great dislocations and property losses as well."

But the attitude of the AEC remained the same. In a 1972 statement to Congress the agency complained of the "widespread lack of knowledge of both the excellent safety record of the nuclear power industry and the extreme efforts, unprecedented in any other industry, to assure that nuclear plants are designed, constructed, and operated with the highest attention to public and employee safety."

About this time the AEC initiated another study of reactor safety. They hired Dr. Norman Rasmussen, professor of nuclear engineering at the Massachusetts Institute of Technology (MIT), to direct a three-year, $4 million study of reactor safety. Rasmussen proposed that the study be made at MIT, but the commission insisted it be done at AEC headquarters in Maryland.

Rasmussen and a staff of about fifty worked closely with the AEC. A draft of the study was released in August 1974 and widely distributed to government agencies, utilities, environmental organizations and interested individuals who were invited to suggest changes for the final report. Many of the comments received were highly critical. Nevertheless the final report, published in October 1975, was only changed in trivial ways.

The nine-volume Reactor Safety Study (often called the Rasmussen Report) has been called a biased whitewash and further evidence that nuclear development should be slowed down or halted. It has also been hailed as an independent and definitive study and a green light for full-speed nuclear development. It remains the most ambitious study attempted so far.

The Rasmussen Report limited itself to accident risks (excluding sabotage) at commercial nuclear reactors in the United States. It did not evaluate dangers throughout the nuclear fuel cycle, or risks at breeder reactors. Unlike the first two safety studies, it did not concentrate on the worst that could happen, but emphasized analysis of the chances of different kinds of nuclear accidents. Rasmussen and his staff included information on such accident-causing factors as mechanical failures, poor maintenance, and human error. From these data they calculated the chances for accidents, and also possible damage.

The study concluded that chances of a core meltdown were greater than estimated by earlier studies. (With a hundred reactors operating, there would be one meltdown in two hundred years.) However, the study concluded that most core melts would not cause public harm, partly because advance warning would allow several hours for evacuation of people near a power plant.

The worst possible accident, according to the final Rasmussen Report, would cause 3,300 early deaths, 45,000 injuries, and $14 billion in property damage. (All of these figures represent increases—changes made by Rasmussen and his staff—over the estimates given in their draft report.) However, the overall tone of the report and its summary was optimistic and emphasized that chances of any serious nuclear accident were extremely small. Precisely, the chances of a person's being killed as a result of a reactor accident were one in five billion—about the same risk as being hit by a meteorite. By comparison, a person is much more likely to die by lightning (a one in two million chance) or in an airplane crash (one in a hundred thousand).

Among the criticisms of the study were those from a panel of twelve physicists who represented the American Physical Society. They pointed out miscalculations which resulted in underestimations of human deaths. The panel also expressed doubts about the safety-analysis technique used to estimate the probability of accidents. This technique was also criticized by the Sierra Club and the Union of Concerned Scientists in a joint study of the Rasmussen Report draft. The critics pointed out that this estimating technique was once used by the aerospace industry but had been abandoned as unsatisfactory for estimating safety or reliability. When the critics applied this technique to one kind of reactor accident, they found that the accident's probability of happening was extremely small—one in a billion billion. Yet the mishap had already occurred, in 1970, at an Illinois reactor.

The final Rasmussen Report was also criticized by the

Environmental Protection Agency. Its 1976 review stated that the study seriously underestimated the number of likely deaths and illnesses that would result from a reactor accident. In 1977 a Ford Foundation study concluded that the Rasmussen Report's estimates of accident hazards might be five hundred times too low, or too high.

Finally, the Nuclear Regulatory Commission (NRC), successor to the AEC, appointed a panel to evaluate the Rasmussen Report. The panel's 1978 judgment was that the report was a failure, and early in 1979 the NRC itself called the report unreliable. Thus, the government found itself with no credible estimate of the risk to human safety and health presented by nuclear reactors.

Other revelations made during the 1970s caused additional worry about reactor safety. When a LOCA occurs, a reactor's Emergency Core-Cooling Systems (ECCS) are designed to prevent a meltdown. The AEC and utilities had long boasted about the ECCS, a key element in reactor safety. But a Harvard student who was studying the economics of the nuclear-power industry discovered some obscure reports that raised serious doubts about the emergency cooling systems of reactors. The reports indicated that the only "proof" that the ECCS would work was computer studies. There was no direct evidence, except for some small-scale tests, conducted in 1970, using nine-inch-high "reactors." (The core heat was provided by electric heaters.) Five tests were tried and in each one the miniature ECCS failed. The AEC emphasized that these small-scale tests did not really apply to large, modern reactors. Still, they were the *only* tests.

The Union of Concerned Scientists called for a suspension of reactor licensing until safety systems could be proved reliable. Protests mounted, and the AEC held hearings on this matter for nearly a year. From testimony and cross-examination of witnesses, and from AEC documents, it was clear that some AEC safety researchers also doubted the reliability of the ECCS.

Some long-delayed safety tests were begun in December, 1978, in an Idaho test reactor that is one-sixtieth the size of a modern reactor. The studies involve actual nuclear fuel, and will provide important information about the ECCS. Because of its design and small size, however, the test reactor can't be used to check some kinds of potential accident situations. Some opponents of nuclear power argue that tests in a full-size reactor are vital, though hazardous.

Years of accumulated abuses had finally brought the Atomic Energy Commission to an end in January 1975. Its licensing and regulating responsibilities were assigned to the Nuclear Regulatory Commission (NRC). Its research and development responsibilities went to the Energy Research and Development Administration (ERDA), which in 1977 became part of the new Department of Energy. Long-time critics of the AEC hailed its death, and the separation of its functions. But most AEC employees were transferred to the two new agencies. Their pronuclear zeal did not die with the AEC.

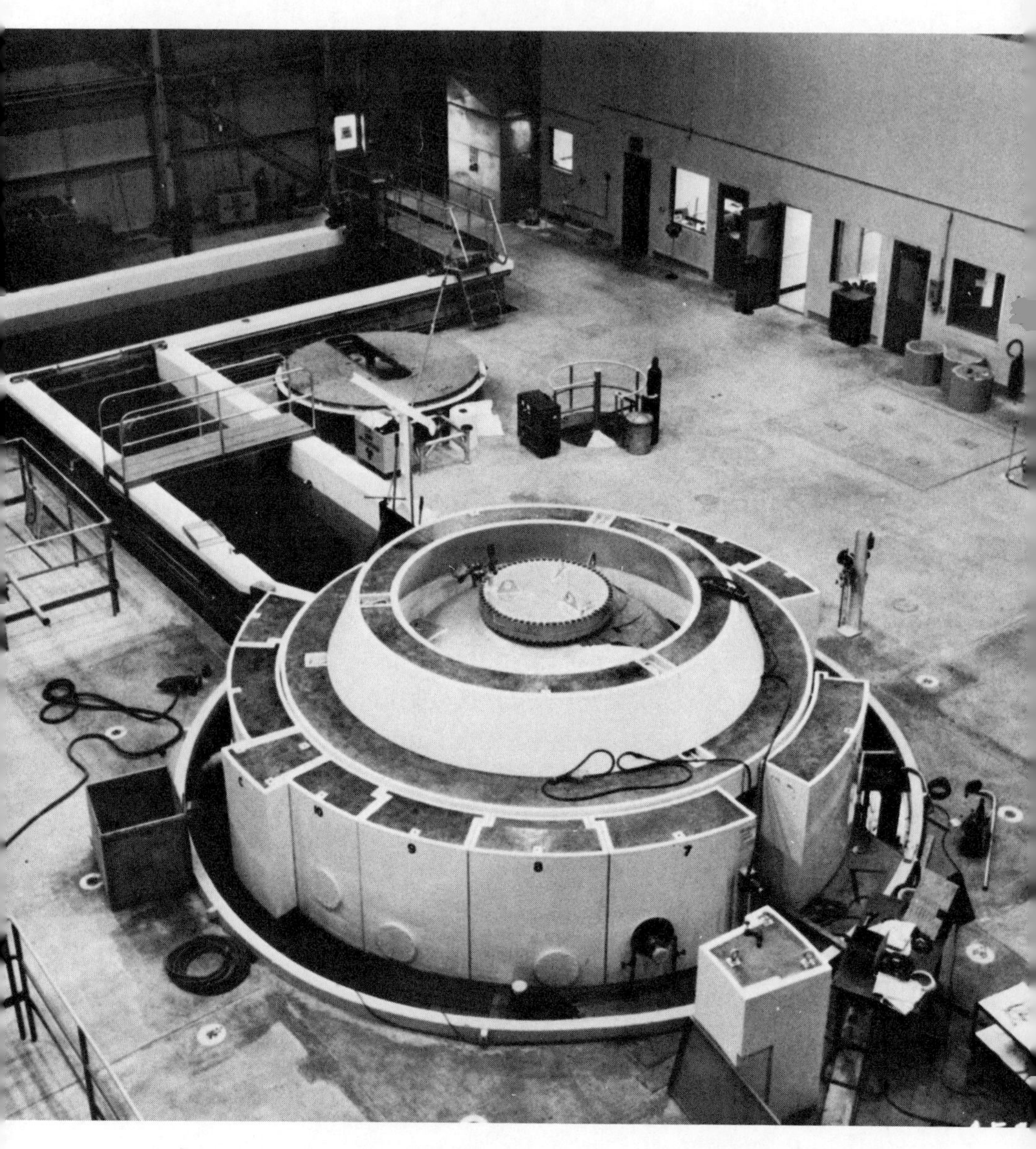

Tests of emergency cooling systems finally got underway in late 1978 at the National Engineering Laboratory in Idaho.

THE BROWNS FERRY INCIDENT

On March 22, 1975, the two large reactors at Browns Ferry, Alabama, were operating and generating 2,200 megawatts of electricity (one megawatt equals 1,000 kilowatts). Beneath the plant's control center, electricians were trying to seal air leaks among a complex array of electrical cables—the electrical system that controlled the two reactors.

The leaks were being sealed by stuffing strips of polyurethane foam among the cables. To test for air leaks, the electricians held a lighted candle near the plastic foam to see whether the flame flickered. Some of the foam caught on fire. Three chemical extinguishers failed to put it out. After fifteen minutes a fire alarm was sounded. Meanwhile, the fire was spreading and destroying cables and affecting the plant's electrical systems.

In the control room, lights on the control panel flashed on and off and alarms sounded. Both units of Browns Ferry were shut down by insertion of control rods, but the possibility of a core meltdown remained. Just twenty-five minutes after the fire began, it had destroyed the electrical system used to control the emergency cooling system of one reactor. A small pump was used to keep water covering the fuel core. Control of most of the ECCS of the second reactor was lost an hour later. A meltdown was prevented because remaining subsystems of the ECCS worked.

The fire continued. It burned for six hours, knocking out more electrical systems and the plant's telephones.

Three reactors are located beside Wheeler Lake at Browns Ferry, Alabama.

Firefighters turned on an extinguishing system specially designed to put out a cable-room fire. It failed. Firemen from nearby Athens, Alabama, eventually doused the blaze with water.

The fire had been confined to one concrete-walled room but had destroyed 1,600 cables. Both reactors were shut down for nearly eighteen months. The cost of repairs and buying electricity from other sources totaled more than $300 million.

At Browns Ferry and other nuclear power plants, a feather, not a candle, is now used to test for air leaks in cable rooms.

Various conclusions can be drawn from the Browns Ferry incident, depending on one's viewpoint. To some, the events at Browns Ferry were dramatic evidence that nuclear reactors have "defense in depth." Even though some emergency systems failed, the reactors were shut down with no damage to their fuel cores. No radiation was released and no one was injured.

To others, the Browns Ferry incident represents a close call with catastrophe, and further proof that nuclear technology may be too dangerous and unmanageable. One worker with a candle unwittingly revealed serious flaws in the safety systems at one power plant, and brought to light inadequacies in regulation and inspection methods.

The Browns Ferry incident brought the issue of nuclear safety into the public spotlight, but it is only one of many nuclear mishaps that have occurred in the United States and elsewhere.

At Soviet reactors there have been several accidents, radiation releases, and an explosion, though few details have been revealed. In 1966, a small breeder reactor near Detroit, Michigan, suffered a partial core meltdown—the result of several human errors before and during the emergency. There are other incidents, including one at a Vermont reactor where control rod parts were installed upside down, and where on another occasion fission was started while the top of the reactor was left uncovered. Antinuclear groups eagerly publicize such incidents, which could, theoretically, lead to a serious accident. They also remind us that fallible humans are in charge, and can make mistakes. Pronuclear groups look at the same events and say, "There was no disaster and no harm to the public. This proves again that our extraordinary safety efforts work."

SOME "INSIDERS" RESIGN

For many years the nuclear-power industry and the AEC fended off critics with a "leave it to the experts—we know best" attitude. They could always fall back on the argument that only those in the nuclear-energy field could really judge its safety, and "everyone who really understands it is for it."

This argument was shaken in 1976 when several people "who really understood it" resigned their jobs because they felt that they could no longer support nuclear power. One, named Robert Pollard, had worked more than six years for the AEC and its successor, the NRC. He was project manager for eight reactors that were being built in 1976. Before resigning he also had reviewed the safety design of one reactor operating at Indian Point, New York. He called the reactor "an accident waiting to happen."

Most significant, perhaps, was Pollard's decision that only a dramatic, well-publicized resignation might change attitudes and actions at the NRC. He had concluded that it was futile to improve reactor safety from inside the NRC. As his letter of resignation put it, "The plain fact is that many of the dedicated government employees in the NRC are deeply troubled about the pervasive attitude in the NRC that our most important job is to get the licenses out as quickly as possible and to keep the plants running as long as possible. . . ."

In response, the NRC evaluated Pollard's charges and "found no need for regulatory action." The agency also concluded that the reactor at Indian Point "meets our requirements for safety."

At about the same time, three engineers at General Electric's nuclear division also resigned. General Electric manufactures reactors and other nuclear equipment. These men had been involved in the development of many reactors sold both in the United States and abroad. One engineer said, "We cannot design to cover human error, and I am convinced the safety of nuclear reactors hangs on the human error."

A spokeman for General Electric said that the engineers had presented "no fresh views or arguments but repeated emotional claims."

THE THREE MILE ISLAND INCIDENT

The worst nuclear accident in the nation's history so far occurred in late March, 1979. It happened at one of two reactors on Three Mile Island in the Susquehanna River, south of Harrisburg, Pennsylvania.

For the first day or so there seemed to be nothing extraordinarily serious about the incident, although there was damage to the fuel core and release of some radioactivity into the atmosphere. A utility spokesman said that the emergency was over, and that "there was nothing that was catastrophic or unplanned for."

By the evening of Friday, March 30, however, this optimism had vanished. That night the nuclear dream turned into a nuclear nightmare for millions of people who lived downwind from the crippled reactor. A catastrophic meltdown *was* possible, and the problem *was* unplanned for.

The source of the trouble, never anticipated during decades of nuclear safety studies, was a large bubble of hydrogen gas which formed inside the damaged reactor. If the bubble had expanded or exploded, it might have incapacitated pumps of the cooling system, and also uncovered some of the fuel rods, causing a meltdown. Once the trouble was understood, technicians tried to devise ways to reduce the size of the bubble.

Meanwhile, Pennsylvania's governor, Richard Thornburgh, urged preschool children and pregnant women to leave the area within five miles of the plant (fetuses and young children are most vulnerable to harm from radioactivity). Pennsylvania officials hastily contrived plans for evacuating a million people. About 100,000 fled voluntarily.

The evacuation was not needed. A way was found to gradually reduce the size of the bubble in the reactor. The crisis eased in a few days, although weeks passed before the reactor was completely shut down. Whether it would ever produce electricity again was in doubt; cleanup and repair would probably take several years and might be so costly that the effort wouldn't be made.

The Three Mile Island incident was investigated by Congress, as well as by an eleven-member panel chosen by President Carter. Long before the reactor cooled, however, some facts about the incident were reported.

The accident was the result of several kinds of human error. Some equipment failed. Some devices were improperly designed to cope with the situation. Several errors by the plant's operators contributed greatly to the near-disaster. Good luck had as much to do with averting catastrophe as good planning.

These revelations did not inspire public confidence in the nation's nuclear program. Especially disturbing were the transcripts of emergency meetings by the five commissioners of the NRC. Chairman Joseph M. Hendrie complained about the lack of information available, even on the third day of the crisis. Referring to himself and Governor Thornburgh of Pennsylvania, he said, "We are operating almost totally in the blind; his information is ambig-

Nuclear reactors are operated by humans from control rooms that look like this.

uous, mine is nonexistent and—I don't know—it's like a couple of blind men staggering around making decisions."

Dr. Roger J. Mattson, director of safety for the NRC, said, "Bringing this plant down is risky . . . No plant has ever been in this condition, no plant has ever been tested in this condition, no plant has ever been analyzed in this condition in the history of this program."

In a few days the incident at Three Mile Island did more to lessen public confidence in nuclear power than all the antinuclear demonstrations ever held. The public saw that nuclear experts were surprised and frightened by an accident at one of the nation's newest reactors, equipped with the latest fail-safe devices and subjected to the NRC's strictest safety reviews.

Later revelations dealt further blows to the credibility of nuclear proponents. The United States General Accounting Office reported that the NRC had been very lenient with many utilities over violations of its rules. In a recent year the NRC had found 2,500 violations, yet had imposed only thirteen fines. The amounts fined were well below the maximum allowed by law. Furthermore, it appeared that the NRC had allowed the new reactor at Three Mile Island to be rushed into operation on the next-to-last day of 1978; this enabled the utility to take a $40 million tax deduction for that entire year. Both before and after that date, the plant had been plagued by many equipment failures and shutdowns.

Robert Pollard, formerly of the NRC, said that problems similar to those that touched off the accident had occurred at several other reactors. "If anyone had been paying attention," he said, "Three Mile Island wouldn't have happened." Among those not paying attention were officials of the Babcock & Wilcox Company, builder of the crippled reactor. Eleven months before the near-disaster, a nuclear safety expert had sent the company a detailed warning about the kind of design errors which contributed to the accident. About a month after the accident at Three Mile Island, the NRC ordered changes made in all Babcock & Wilcox plants.

Nuclear proponents continued to point out that safety systems had halted the accident, and that no one had been hurt. However, this last claim might be amended in time. Some scientists believe that even the low levels of radiation released could have harmed people living near the plant. Long-term studies were set up to check the health of the public, and of utility workers who received greater radiation exposure.

The Three Mile Island incident was a major landmark in the history of nuclear development. It had two very different effects. One was a renewed effort to perfect nuclear power plants and their operations, to make them safer. There were calls for increased and more strict government supervision, better valves and gauges, better-trained operators, and—just in case—better emergency evacuation plans.

The second effect was quite different. The accident caused the plant to stop generating electricity, and it began generating fear. Millions of people felt vulnerable to a mysterious, invisible, far-reaching, and awesome danger. The crisis ended, but some of that feeling remained. "Not all the safety assurances in the world," said Pennsylvania's governor, "can erase the awareness of these good people that something out there is powerful and strange and not entirely under control."

The feelings of people who had been antinuclear were strengthened, and millions of others began for the first time to oppose nuclear development. The nuclear enterprise was on trial as never before.

An open-pit uranium mine in Wyoming

3

DANGERS IN THE FUEL CYCLE

Throughout the nuclear fuel cycle there are hazards to workers and often to the public. Nuclear proponents point out that this is true of other fuels, a prime example being coal which pollutes the air and causes black-lung disease in miners. But nuclear fuel has unique dangers that may not be as easily avoided as those of coal.

Workers in poorly ventilated mines and mills inhale uranium dust and radioactive gases (especially radon and its decay products). In the 1940s, studies showed that a large percentage of European uranium miners died of cancer. Dangerous levels of radioactivity were then found in uranium mines in the western United States, but little was done to improve conditions. Finally, in 1967, the United States Department of Labor set radiation-exposure limits for uranium miners. According to the United States Public Health Service, the maximum levels of radioactivity allowed may still be too high, since greater than normal numbers of miners die of cancer.

Milling wastes pose a threat to the general public in some states. Piles of leftovers resembling fine sand, called tailings, are left exposed to wind and water. They have contaminated rivers and community water supplies with radioactive radon, radium, and thorium (which has a half-life of 80,000 years).

Before the dangers of these wastes were known they were used as fill and construction material in Grand Junction, Colorado, the site of a uranium mill. Occupants of hundreds of homes and a school were exposed to dangerous levels of radioactivity. Since the problem was recognized, the federal government has spent about seven million dollars to remove tailings and rebuild homes. By the late 1970s, the job was only half done.

About two tons of tailings were located within Salt Lake City, Utah. In 1978, the human exposure to radon in a Salt Lake City firehouse—built on tailings—was discovered to be seven times greater than that allowed for uranium miners. Overall, there were 140 million tons of tailings in 1978, and as many as a billion tons expected by the year 2000. Unless the tailings are somehow buried, they will release radon to the environment for more than 100,000 years. Environmental groups petitioned the federal government to face this problem. Finally, in 1978, Congress approved a clean-up program. Its effectiveness and cost to the public remain to be seen.

Farther along in the fuel cycle there are other troublesome wastes. At government-run uranium-enrichment plants, large amounts of uranium hexafluoride are left over. This waste has so far been stored temporarily in steel drums and awaits some kind of long-term storage or the development of a future use. The handling of uranium as it is finally prepared for use as reactor fuel also poses some radiation threat to workers, though unfissioned uranium is comparatively harmless next to plutonium, another potential reactor fuel.

During fission, uranium is like coal or other fossil fuels in one way: It produces lots of waste heat. Just one-third of the heat produced by fission is converted to electrical energy. The other two-thirds is wasted. (Fossil-fuel plants are somewhat more efficient, although they waste more than half of the heat energy they produce.)

The waste heat from a nuclear reactor has to go somewhere. Most often it is carried off via water. Utilities call this thermal enrichment; environmentalists call it thermal pollution. This heated water has been proved to have several bad effects on fish in the rivers, lakes, or bays to which it is returned: It may kill the fish directly, prevent their eggs from hatching, or stimulate the growth of disease organisms.

Fish are attracted to the warm water during wintertime. Many may be killed when sucked against screens covering water-intake pipes. Or huge numbers of them may die of thermal shock when a reactor shuts down and the fish are suddenly faced with normal winter temperatures. This has happened several times at New Jersey's Oyster Creek nuclear plant, where brief winter shutdowns have killed hundreds of thousands of menhaden.

Problems like these have caused so much concern that some utilities have been forced to build cooling towers. In the most common type, large amounts of water evaporate into the air. This water must be replaced from a nearby lake or river, but overall the demand for water is much less because most of it is not lost and, once cooled, is used again and again.

Utilities have been exploring the idea of building nuclear plants in clusters or "energy parks." This would put a tremendous strain on a region's water supply and water quality, as ten or more power stations each sucked in a million gallons of water a minute and then returned it with a load of heat. Even if cooling towers were used, there might be bad effects on the region's climate. According to an NRC study, a cluster of twenty nuclear plants would require forty cooling towers five hundred feet tall. They would send water-vapor clouds seven thousand feet into the air, and the clouds would extend as far as fifty miles downwind.

Since large nuclear plants need fifty percent more cooling water than same-sized fossil-fuel plants, a lack of water may keep them out of some regions. In California, the San Joaquin Nuclear Project was planned to be the biggest nuclear plant in the world. But it would have needed twenty billion gallons of cooling water annually in an inland area where there is already sharp competition for scarce water

Near Sacramento, California, a reactor's cooling towers reduce the need for massive amounts of water.

supplies. The project was cancelled. In California the Pacific Ocean is an obvious source of cooling water for reactors, but the danger of earthquakes near the coast has already caused the cancellation of one nuclear project and the closing of a small research reactor.

All things considered, however, problems of water supply and thermal pollution seem rather mild compared to the dangers associated with fission. Here the difference between nuclear energy and other fuels is striking. For all of their bad environmental and health effects, fossil fuels, with the exception of liquified natural gas, do not have a potential for sudden catastrophe. No major accident at a coal-fired plant could cause as much death and damage as such an accident at a nuclear plant. There is concern about the long-range effects of adding large amounts of carbon dioxide to the atmosphere, but we have already developed devices and methods to control the other solid and airborne wastes of fossil-fuel power plants. Many people believe that this is not true of nuclear wastes.

"WHAT DO YOU DO WITH THE STUFF?"

"The stuff" is nuclear wastes. The speaker was the late Vice President Nelson Rockefeller. While governor of New York State, Rockefeller welcomed a nuclear reprocessing plant as a "symbol of imagination and foresight." To lure the world's first commercial nuclear-waste reprocessing plant to New York, the state agreed to assume full responsibility for the wastes, should the plant fail. It closed in 1972.

Up to a half-billion dollars in state or federal tax funds will be needed to dispose of 600,000 gallons of highly radioactive wastes—and the radioactive plant itself.

Next to a reactor accident, the disposal of nuclear wastes is the public's greatest concern. There is wide disagreement on the seriousness of this matter. Former AEC chairperson and governor of Washington, Dixie Lee Ray, said that nuclear waste is "the biggest non-problem we have," one that is "readily solvable."

But the "non-problem" worries people. Dr. Harvey Brooks, dean of the Division of Engineering and Applied Physics at Harvard University, predicted that "should nuclear energy ultimately prove to be socially unacceptable, it will be primarily because of the public perception of the waste disposal problem."

So far the public perception is that wastes are extremely dangerous for a long, long time, and there is no proof that they can be stored safely.

Each year about one-fifth to one-third of a reactor's fuel is replaced. The spent fuel is left in its metal rods. In the United States, nearly all of the commercial nuclear waste is now stored in pools of water at power plants. Radiation gives the cooling water a bluish glow. Storage in cooling water allows some short-lived radioactive materials to decay until harmless. One such element is iodine-131, with a half-life of eight days.

In 1977 about 2,500 metric tons (a metric ton is 2,204.6 pounds) of spent fuel were stored this way. By the year 2000, when many more reactors may be operating, the

The water glows as short-lived radiation is given off by spent fuel rods stored at reactor sites.

accumulated wastes may have grown to 25,000 metric tons, which is about 40 to 60 million gallons.

Military wastes add to this total. They have been accumulating since 1945 and in 1977 amounted to about 75 million gallons. Two-thirds of this waste is stored as liquid in underground steel tanks near Hanford, Washington. In 1973, about 115,000 gallons of radioactive wastes leaked from one tank into the surrounding soil, but did not reach groundwater. The tank leaked for fifty-one days before discovery. Like many other nuclear mishaps, this one was mostly a result of carelessness.

The total amounts of nuclear wastes may seem large, but the actual volume is not great. A typical large reactor produces from two to three cubic meters of waste annually—an amount that will fit under a dining room table, as one nuclear advocate is fond of saying. In contrast, a large coal-fired plant equipped with air-pollution devices would collect an estimated 150,000 metric tons (10 billion gallons) of sulfur dioxide wastes annually.

It is the *content,* not volume, of nuclear wastes that really matters. Spent reactor fuel can be divided into two categories. One is fission products with fairly short half-lives, such as strontium-90 and cesium-137. With half-lives of thirty years or less, elements like these will be harmless in about six hundred years.

The second group is called transuranics and it includes plutonium-239, neptunium-237, and americium-243, all with very long half-lives. Almost a quarter-million years must pass before plutonium decays through ten half-lives;

throughout that whole span it can cause cancer and genetic damage. The transuranic situation gets even worse with time. As americium-243 decays it becomes plutonium-239. So a quantity of transuranic wastes becomes more dangerous after 20,000 years because by then the amount of plutonium-239 has increased.

If all of these wastes are left together, they must be safely stored out of human contact for a half-million years. However, if only the fission products have to be stored, the waste problem is much less troublesome—a matter of storage for "only" seven hundred to a thousand years. This would be the case if reprocessing could remove all of the transuranics. However, the reprocessing method developed so far does not do that.

Throughout the brief history of commercial nuclear power, people expected that reprocessing would be a major part of the fuel cycle. People in the business do not think of spent fuel as waste, since it contains fissionable nuclei. They assumed that transuranics would be recycled. Uranium would be sent to an enrichment plant to be readied as fuel for water-cooled reactors. And plutonium would be used directly as fuel in breeder reactors.

These expectations may not be met. The cost of reprocessing has proved to be so great that the small amounts of uranium fuel recovered are no bargain. As for breeder reactors, they have developed more slowly than expected in the United States. They may never be used here because of concern about plutonium. In addition to being one of the most deadly substances on earth, plutonium in rather

The Nuclear Fuels Services Reprocessing Plant in New York State, an economic failure, is itself a radioactive disposal problem.

small amounts can be made into a nuclear weapon. In the late 1970s many citizens were questioning whether the United States should get into the business of producing pure plutonium, either in breeder reactors or through reprocessing of spent fuel.

The only reprocessing method in use was specifically invented, in the early 1950s, to produce plutonium for nuclear weapons. In fact, six of the seven nations that have tested nuclear weapons used this process to get the plutonium they needed.

In 1977, the United States government began a series of steps to halt the spread of nuclear materials and weap-

"Remember The Good Old Days When We Only Worried About Russia Getting One?"

ons. One step was to defer indefinitely fuel reprocessing in the United States—a move supported by environmental and arms-control groups, and opposed by the nuclear-power industry. The future of a half-finished reprocessing plant in South Carolina was left in doubt.

The example set by the United States was not followed by Britain, France, Germany, and Japan. They continued to reprocess spent fuel, or to plan reprocessing plants, though public opposition was growing in these nations, too. Canada deferred a decision on whether to reprocess nuclear fuel, and stored its spent fuel in aboveground concrete silos that were considered safe for fifty to a hundred years.

Reprocessing may yet be possible without adding to the earth's supply of pure plutonium. New methods may recover the fuel value without separating plutonium. In any event, with the matter of reprocessing in question, the problem of nuclear-waste disposal may become more troublesome. Researchers can no longer assume that plutonium and other transuranics will be mostly separated from the fission byproducts with shorter half-lives. Separation of the transuranics might still be done, especially if studies show that this step is vital for safe storage. Reprocessing could be done by the government, and plutonium and other transuranics destroyed or changed in special reactors, if necessary. This is considered to be a very expensive, last-resort solution.

The reprocessing issue may be unresolved for decades. Meanwhile, it seems that we may need safe storage of spent nuclear fuel for a half-million years. This is much longer than the entire recorded history of humankind. Can we devise a way to lock up dangerous radioactivity that long?

"Of course we can," said the AEC in 1971. News releases announced the selection of a permanent disposal site for nuclear wastes, deep underground in a Kansas salt formation. Congress was assured that the site had been carefully chosen and investigated; there would be "no significant impact on the environment."

Later that same year the AEC abandoned the site. An active salt mine was found less than a half mile from the proposed permanent disposal site. Water pumped into a shaft by miners had been "lost" in the salt formation, and this raised serious questions about the safety of the site.

Both the actions and inactions of the AEC during the agency's twenty-nine-year life helped arouse public concern about nuclear wastes. Little research money was spent on the "back end" of the fuel cycle. This has changed dramatically. In 1978, the Department of Energy budget called for $116 million to be spent on nuclear waste studies, an increase of almost a hundred million dollars over the years 1975 and 1976.

By that time the nuclear-energy business was clearly in trouble, with the issue of waste disposal a major source of growing public opposition. In 1976 two states (California and Iowa) passed laws which forbid approval of new nuclear plants unless their legislatures are convinced that disposal methods for nuclear wastes exist. In a 1977 hearing on this matter, the California Energy Resources and Development Commission declared the federal waste management "a disaster area." One commissioner said, "The government is anticipating a 1,000-reactor economy without really being able to tell you how they're going to deal with the waste."

In 1977, the United States General Accounting Office (GAO) noted that ERDA had begun a program to demonstrate, by the mid-1980s, the feasibility of storing nuclear wastes deep underground. The GAO's evaluation of the effort concluded, "Not only has progress been negligible to date, but future program goals are overly optimistic because the Energy Research and Development Administration faces many unresolved social, regulatory, and geological obstacles."

Later that year, the Council on Environmental Quality, which advises the President on environmental matters, recommended that licensing of nuclear plants cease if safe methods of waste storage were not proven soon. Realistically, however, a deep underground storage area would not be ready until 1988 at the earliest.

Salt beds are considered by many scientists to be the safest long-term disposal sites. They are among the most stable geologic formations known. Since salt dissolves in water, the existence of a large salt deposit means that little water is present—a vital condition if nuclear wastes are to stay where they are put. Salt is also "plastic"—it tends to heal cracks or other damage caused by radiation heat or earthquakes. This could be important, considering the storage time needed. A lot can happen to the earth's crust in a half-million years. The next ice age, for example, is predicted to begin "soon," in the next 10,000 years.

Another reason for deep burial in salt or perhaps rock is that the disposal site, once permanently sealed, must be inaccessible. We cannot expect generations far in the future to guard our nuclear wastes. Every possible care must be taken to keep the wastes from being discovered accidentally. Wastes must be buried very deep in a place where people would be unlikely to drill for valuable ores.

In an abandoned salt mine, the use of salt formations for storage of nuclear wastes is investigated.

Some detailed plans for storage in salt formations have been worked out, with the assumption, however, that most transuranics would first be removed by reprocessing. According to Dr. Bernard Cohen, professor of physics at the University of Pittsburgh, solidified wastes would be encased in glass cylinders, with each cylinder then sealed inside a stainless-steel canister. It would be best not to bury the canisters for about ten years, to allow the wastes to emit most of their heat. They would then be placed in a storage area 600 meters (a third of a mile) underground. The shaft to the storage area would be sealed so that no radiation could escape.

Salt formations are not completely dry, and Dr. Cohen assumes that heat from the canisters will cause some water to collect around them and that the steel will eventually corrode away. Small amounts of water are not a problem, in his opinion. Only water that could circulate out of the salt to the surface would be dangerous, and this problem would be avoided by choosing a huge salt deposit where circulating groundwater does not exist.

Dr. Cohen is convinced that nuclear wastes can be stored safely. One benefit of nuclear power, he believes, will be a reduction in radiation damage to humans. According to his reasoning, the natural background radiation on earth will lessen as more and more uranium at or near the surface is mined and fissioned, with the wastes then safely buried. "Thus," he concludes, "on any long time scale nuclear power must be viewed as a means of cleansing the earth of radioactivity."

This view is disputed by other scientists who point out that the very first steps in the nuclear fuel cycle, especially the creation of piles of tailings, have added to the radiation dangers on earth. Dr. Cohen believes that this problem is solvable and temporary.

Revelations in 1978 raised doubts about such disposal schemes. There seems to be more water in salt crystals than was earlier believed; a panel of geologists reported that "we are only just learning about the problem of

Radioactive wastes have leaked from steel storage tanks near Hanford, Washington.

water in salt beds." And a National Academy of Sciences study concluded that glass, long considered the ideal container for nuclear wastes, is less stable, particularly at high temperatures, than other substances, especially ceramics.

Disposal sites other than salt beds have been suggested or are under investigation. They include outer space, Antarctica, and the ocean floor, where clay deposits have been undisturbed for hundreds of millions of years. Nuclear scientists in Canada are investigating a rock formation that has changed little in a billion years or more. The West German government has selected a salt formation for deep storage of its nuclear wastes.

Because of earthquake dangers or other geological factors, some nations, such as Peru, Indonesia, and Japan, may have no safe sites for long-term storage. International cooperation will be needed, since the consequences of a storage failure might be global. The United States government has offered to accept and store spent fuel from other nations as part of its nuclear nonproliferation policy.

In 1977, the federal government proposed to charge utilities a one-time fee and take responsibility for their spent fuel. The annual fee for a large reactor could reach $3 million, and utilities would pass this cost on to their customers.

"Who pays?" is a question raised about another kind of nuclear waste: used reactors. They are licensed to operate for forty years. Changes in nuclear technology will make them obsolete, so most or all old reactors will be shut down for good. But they won't be ordinary defunct power plants. The reactor vessel, lots of piping, and perhaps other equipment and structures will be radioactive. The power plant will have to be "decommissioned."

One decommissioning method is to lock and guard the plant until the radioactivity inside decays to harmless levels. This could take two hundred years or more. Radioactive parts could also be left in place but entombed in concrete, so that less guarding would be needed. Or the radioactive parts and structures could be taken apart. No long-term guarding would be needed, but the dismantled parts would have to be stored safely, probably deep underground.

'Just Keep Driving Around—We May Come Up With A Solution Yet'

This problem was not foreseen by the AEC, and nuclear power plants are not designed to be easily decommissioned. By the year 2000 there will be more than a hundred inactive but radioactive power plants worldwide. The NRC has estimated dismantling costs for a large power plant at $36 to $60 million. Industry estimates are higher—up to $150 million. The cost and the difficulty might be even greater, depending on yet unknown effects of fission on reactor vessels. According to a study by Dr. Martin Resnikoff, professor of physics, State University of New York at Buffalo, part of the inner surface of reactor vessels might be converted to the element nickel-59, which has a half-life of 80,000 years.

Even without this possible problem, decommissioning is a neglected matter that will have to be dealt with, and paid for. Who will pay? Utility customers? Taxpayers? This economic and political problem must be faced soon.

The back end of the nuclear fuel cycle also raises moral and ethical questions. By burying our radioactive wastes we reduce the danger to us, but is it fair to burden future generations with potential risks from nuclear wastes?

Nuclear advocates believe that the danger of the wastes—now and in the future—has been greatly exaggerated. Besides, they argue, we have chosen to put other burdens on future generations, for example, by burning up so much of the earth's fossil fuels in the past century. We can ease their burden by giving our descendants the gift of nuclear energy.

Judging from the growing opposition to nuclear development, many people question the value of this gift.

SCIENCE
Nuclear Waste Disposal: Can the
Nuclear Power: Hard Times
and a Questioning Congress
How I Designed an
A-Bomb My Junior Year
At Princeton
Nuclear Power Debate:
Signing Up the Pros and Cons
Over Testimony on Lost Uranium
OUR NATIONAL
DEATH WISH
Is Nuclear Too Costly?
Expenses
Soar as
Demand
Softens
Nuclear Licensing: Promised Reform
U.S. Move to Sway California Nuclear Vote
74 NUCLEAR PLANTS
TOLD TO STRENGTHEN
ANTITERRORIST GUARD
Regulatory Agency Concerned Over
a Possible Theft of Plutonium
or Sabotage of Reactors
THE NEW YORK TIMES, SUNDAY, JULY 23, 1978
Study Projects Low Risk From Radioactive Cargoes in Urban Areas
By MATTHEW L. WALD
A Vote This Week
California's
Case Study
In Nuclear
Politics
Nuclear Weapons
International Stability
SEABROOK SUSPENSION
PRAISED AND ASSAILED
Closed nuclear
awaits decision

4

COSTS, BREEDERS, AND BOMBS

"We who are alive today owe our descendants a source of cheap and abundant energy. The only such source we can now guarantee is nuclear fission."

Dr. Bernard Cohen's opinion, expressed in 1977, represents a cherished notion of nuclear advocates: that nuclear power is cheap. Like the claims of "safe" and "clean," however, "cheap" is also being questioned.

The costs of an electric generating plant can be divided into three parts: capital, fuel, and operating expenses. Capital costs include construction expenses and the interest charged on money borrowed for this purpose. Nuclear power plants have higher capital costs than fossil-fuel plants, but nuclear plants are supposed to be cheaper in the long run because their fuel is cheaper. But the capital costs of new nuclear plants have risen dramatically. Uranium prices have also soared, and the difference in fuel costs narrowed.

A large nuclear plant built in 1968 cost $200 million;

the same-sized plant built in 1978 cost $1 billion. An economic study released in 1975 found that "the capital costs of large light-water reactors show no signs of stabilizing and, indeed, are apparently still climbing at alarming rates." This study, conducted by economists from Harvard and MIT, concluded that it would be a great mistake for utilities to assume that the economic advantage of nuclear power over fossil fuels would be permanent. Indeed, in some parts of the United States it had already disappeared.

The rising costs of nuclear power were one reason for a sharp drop in reactor sales. Only four were ordered in 1977, and none in 1978. But there was another reason: people were using less electricity. From 1940 to 1973, electricity use in the United States had grown at the average rate of seven percent annually. Then the cost of electricity rose sharply, and people used less. Utilities found that they could cancel or defer orders for new power plants. They also had time to reassess the long-term costs of coal power versus nuclear power.

One disadvantage of nuclear power is the time needed to build a nuclear plant, from eight to twelve years compared with six years for a coal plant. This has a direct effect on a plant's capital costs, since a utility has to pay finance charges during construction but the plant earns no money then.

Reactors have been delayed for many reasons, including late deliveries of equipment, court challenges by citizen groups, and changes in safety features ordered by the AEC or NRC. Proponents of nuclear power often blame environmentalists for reactor delays. But modern reactors are much larger than those of a decade ago and take longer to build. Also, utilities deliberately slowed work on projects because demand for power was much less than they had once expected.

During the mid-1970s, from two to three-and-a-half years were needed for a utility to get a construction permit from the NRC. So nuclear proponents sought to influence Congress to set up a speedier process of reactor-licensing. After the Three Mile Island incident, however, the public's concern was safety, not speed. Those people who wanted any new reactors at all wanted tougher safeguards against disaster. Stricter regulations and more elaborate safety systems seem likely to add to construction time, and would definitely add to costs.

Besides the great and growing capital costs of nuclear plants, their reliability has been a disappointment. As a writer for the *Wall Street Journal* put it, "Their unreliability is becoming one of their most dependable features."

A power plant is designed and built to produce a certain amount of electricity. If it could generate that amount twenty-four hours a day all year, it would have a hundred-percent capacity factor. But this ideal is never achieved. At some seasons or times of day the demand for electricity lessens, so the power plant operates below full capacity. The capacity factor of nuclear plants is also reduced when they shut down for refueling. And nuclear reactors have been plagued with unscheduled repairs and shutdowns.

Some kinds of repairs at a nuclear plant are especially troublesome because each worker is allowed only a certain amount of exposure to radiation. Because of this factor, a vital pipe repair at one reactor took seven months and the services of 700 men. A similar job at a nonnuclear plant could have been done in two weeks by 25 men.

The AEC used to assume that a nuclear plant would produce electricity at eighty percent of its capacity. During regulatory hearings in 1976 and 1977, utilities claimed that their proposed reactors would produce electricity at about seventy-five percent of capacity. But the actual performance of most reactors is well below these figures.

A 1976 study by the Council of Economic Priorities compared capacity factors of nuclear and coal power plants. The comparison was between 38 nuclear plants which came into operation from 1968 to 1974 and 250 coal plants which operated from 1961 to 1973. The average capacity factor of nuclear plants was 59.3 percent; of coal plants, 66.9 percent. In 1977, the council published an updated study, based on 48 reactors, which reported that the nuclear capacity factor had declined to 57.5 percent. It appeared that the performance of reactors was not improving with age, as nuclear advocates had predicted.

Late in 1977, the director of this study, economist Charles Komanoff, was asked to testify before the House Subcommittee on Environment, Energy, and Natural Re-

The mining and burning of coal for electricity pose environmental problems quite different from those of nuclear energy.

sources. Comparing the capacities of large nuclear plants (the kinds utilities had been ordering and building) with medium-sized coal plants, Komanoff concluded that electricity from coal would be cheaper in all seven geographical regions of the United States. According to his figures, the advantage would be slight in New England because of the cost of transporting coal there. In other regions, coal-generated electricity would be up to twenty percent cheaper than nuclear.

These findings were disputed by utility executives and government witnesses. An ERDA spokesman said that coal was cheaper in the West, nuclear was cheaper in the East, and "in the Midwest, it's a toss-up." People in the nuclear business also argued that it was too early for such comparisons, and that the capacity factor of nuclear plants would improve with time and experience.

The findings of the Council on Economic Priorities, as well as those of other studies, raised questions about the size of power plants that ought to be built. In the late 1960s, utilities began ordering and building large plants that produce 1,000 megawatts or more. They ordered big power plants in the belief they would be more reliable and therefore more economical. So far this has not been true. The Council on Economic Priorities found that the most reliable plants—both coal and nuclear—were of the smaller, 600-megawatt type. The council's study suggested that "A return to smaller unit sizes (400 to 800 megawatts) could increase nuclear power's competitiveness with coal."

The Three Mile Island incident was expected to have an important economic effect. Soon after the accident, a pro-nuclear economist said, "What the public is likely to do is increase the pressure for more and more safeguards. The economics of coal and nuclear are close enough now that if safety factors alone are shifted enough, we'd all certainly change our view that nuclear is cheaper." But some nuclear advocates continued to speak optimistically of a future with reduced construction times, improved reliability, and energy that would still prove to be cheaper than that generated from coal.

One economic uncertainty facing the nuclear industry is fuel supply and cost. From 1973 to 1977, uranium prices rose from $6 a pound to about $42 a pound. Fuel costs are never expected to return to their former low levels. If uranium prices rise faster than those of fossil fuels for several years, fuel cost could become a disadvantage for nuclear power.

The big increase in uranium prices caused economic troubles for utilities and especially for the Westinghouse Electric Corporation. The corporation was obliged by contract to provide its utility customers with uranium at $9.50 a pound. But Westinghouse didn't own enough fuel to meet its obligations and was faced with the prospect of paying $40 a pound to uranium producers, then getting no more than $9.50 a pound from its customers. Rather than take a huge financial loss, Westinghouse chose to break its contracts. Twenty-seven utilities then sued Westinghouse, saying that the corporation had to fulfill its commitment even if it lost money. In 1976, Westinghouse sued uranium producers in the United States and abroad, charging them with illegally raising and fixing uranium prices.

Up to a point, the rise in uranium prices was welcomed in the nuclear business. For some time the price had been so low that mining companies had little incentive to develop mines or to search for new reserves. Higher prices stimulated a new record in uranium drilling in 1976. Some people feared that exploration and mining would slow again if the market price of uranium dropped below $30 a pound.

Estimates of uranium supplies in the United States vary, but most experts agree that at least 1.8 million metric tons of uranium fuel can be counted on. This is enough for the life span of 350 to 390 large reactors. Double this amount of uranium may be available. Uranium can also be found in Canada, Australia, France, and South Africa. Utilities in the United States can and do use imported uranium, but they may not be able to count on foreign sources; there are growing numbers of nations with reactors competing for the world's uranium.

In the United States, there is wide disagreement about uranium supplies, and even about ways of estimating them. People opposed to proliferation of nuclear weapons and to breeder reactors argue that domestic uranium supplies are large. Therefore, they say we have no immediate need to reprocess spent fuel for plutonium or to develop breeders.

Drilling rigs explore for uranium in Wyoming.

The nuclear industry and its supporters in government tend to be pessimistic about uranium supplies. They claim that domestic uranium won't last much beyond 1990. In fact, some predict a shortage of uranium as early as 1985, with only enough fuel for the plants in operation then; no new plants could be built for lack of fuel. Therefore, they say that we need to reprocess spent fuel and develop breeder reactors as soon as possible.

BREEDING NUCLEAR FUEL

Nuclear proponents have always believed that the present water-cooled reactors would be just a first step in nuclear development. The second step would be breeder reactors. Since the world's supply of fissionable uranium is limited, the future of energy from uranium ultimately depends on the development of *some* kind of breeder.

The idea of breeding nuclear fuel has been around ever since people began to understand nuclear fission. In fact, the first electricity generated by fission came from an experimental breeder reactor in 1951. In the early 1960s the AEC selected one kind of breeder—called the Liquid Metal Fast Breeder Reactor (LMFBR). This breeder is called fast because of the speed of neutrons within its reactor. Its coolant is not water, which would slow the neutrons, but a liquid metal, sodium.

Several AEC studies seemed to justify the huge sums spent on breeder research. The studies tended to underestimate uranium supplies and exaggerate future electrical needs. The agency proposed that a demonstration breeder be built near the Clinch River in eastern Tennessee. Its cost was first estimated to be $400 million; by 1976 the estimate had passed $2 billion.

In 1971, President Richard Nixon declared that "our best

A drawing of the proposed Clinch River demonstration breeder

hope for meeting the nation's growing demand for economical clean energy lies with the fast breeder reactor." That same year Nixon also said, "Now, don't ask me what a breeder reactor is . . . unless you are one of those Ph.D.'s, you won't understand it either."

Six years later President Carter vetoed funds for the Clinch River breeder. He called it "a large and unnecessarily expensive project which, when completed, would be technically obsolete and economically unsound." President Carter favored a continuation of nuclear development, but had taken a strong stand against the spread of nuclear weapons. He said that funds for the Liquid Metal Fast Breeder Reactor could be better spent on other energy programs, including a search for ways to "breed" nuclear fuel without also producing bomb ingredients.

The revival of some kind of breeder reactor remained a possibility, however, because it was favored by powerful political forces. At stake were billions of dollars of potential sales for nuclear businesses. Also, the Soviet Union, Britian, France, and Japan had already built prototype breeders. France and Japan had little coal or oil, and believed that breeders would help make them less dependent on imported oil. Whether the United States joined or not, other nations seemed headed down a new path of nuclear development.

BREEDING NUCLEAR WEAPONS

The fuel of a Liquid Metal Fast Breeder Reactor is plutonium and uranium-238—the most plentiful form of uranium in nature. Uranium-238 is virtually useless as fuel in ordinary reactors. In a fast breeder, however, the neutrons from fissioning plutonium are able to travel fast. They sustain a chain reaction, and excess neutrons escape from the reactor core and are absorbed by a surrounding blanket of uranium-238 atoms. Adding a neutron to a uranium-238 atom converts it to plutonium-239. In this way, some of the uranium-238 is converted into fissionable fuel.

Aside from multiplying the energy content of uranium resources at least sixty times, a fast breeder has other ad-

iquid Metal Fast Breeder Reactor (LMFBR)

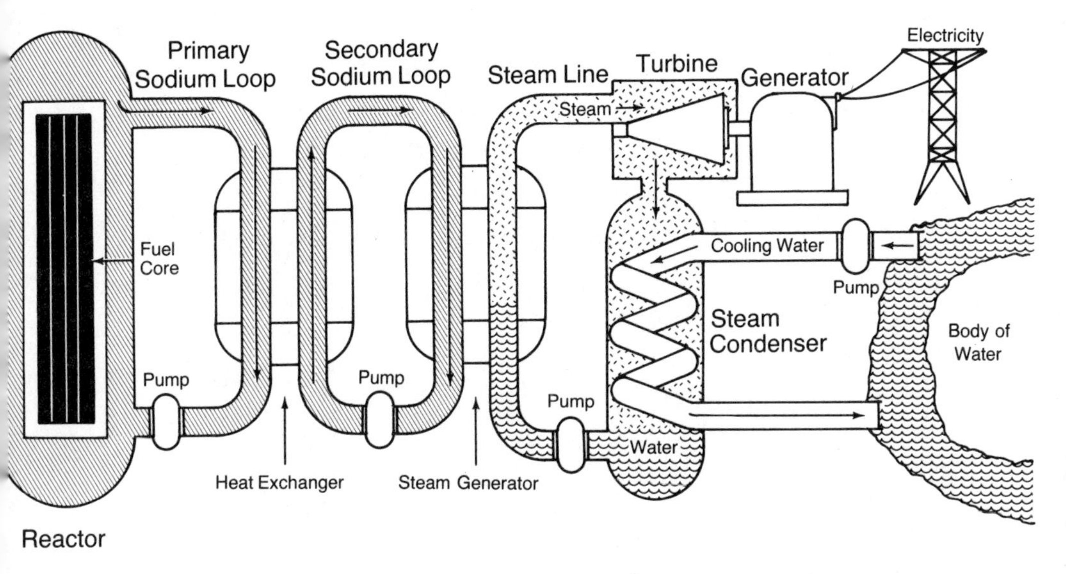

The basic parts of a breeder reactor and electric power plant

vantages. It uses fuel more efficiently than present reactors, that is, it wastes much less heat energy than pressurized water reactors. For example, in its first two years of operation the French prototype breeder, called the Phenix, converted forty-three percent of its heat to electricity.

A fast breeder can operate near 1,000°F (538°C) without the high pressures found in water-cooled reactors, virtually eliminating the chance of a loss-of-coolant accident occurring.

Breeders have disadvantages, too. No large commercial breeders have yet been built, but their capital costs are expected to be very high. Also, the sodium coolant burns explosively on contact with air or water. Extraordinary care must be taken to prevent this. Though loss-of-coolant accidents are not likely, meltdowns and other accidents can occur. A partial core meltdown happened at the only commercial breeder ever built in the United States—the Enrico Fermi plant near Detroit.

A bomblike explosion is also possible. In publications about the LMFBR the AEC tried its best not to use terms like "bomb" or "nuclear explosion." The agency referred to a possible "destructive nuclear excursion" or "disruptive energy release."

The Enrico Fermi breeder, which produced the first nuclear-generated electricity in the United States, was abandoned after a partial meltdown.

However remote the chances of a breeder explosion, the possibility is frightening. A 1,000-megawatt breeder's core would contain 50 metric tons of plutonium and uranium, and 40,000 cubic feet of radioactive molten sodium. A large explosion could hurl much of this material into the air. The plutonium would probably be in tiny airborne particles, its most deadly form.

A fast breeder's fuel cycle would also have some advantages and disadvantages. Since uranium-238 would be used in breeders, much less material would be left over at uranium mines and mills. In fact, existing piles of tailings would be a prime source of uranium for breeders. The expensive fuel-enrichment stage would be unnecessary.

At the back end of the fast-breeder fuel cycle, however, there are many problems. Reprocessing of spent fuel would be vital, in order to recover the plutonium created. But the spent fuel from breeders contains more radioactivity and gives off more heat than spent fuel from nonbreeders. Even those nations that are most advanced in breeder development—France and Britain—have much to learn about the difficulties and costs of reprocessing breeder fuel.

Further development of the LMFBR means the beginning of a "plutonium economy" in which plutonium fuel will be as vital as fossil fuels are today. Some people believe that this will lead to disaster. Full-scale operation of breeders will bring national and international traffic in plutonium. Eventually, it is feared, a human error will cause an accident at a power plant, or during reprocessing or shipping. Another possibility: Terrorists could seize a plutonium shipment, make a crude bomb, and threaten to explode it unless their demands were met.

A nuclear physicist who once designed bombs claims that it is possible for a person to design and build a crude fission bomb from 10 kilograms (22 pounds) of plutonium, using information from publications that are available to the public. Such a bomb would be small enough to carry in a car, yet would have the explosive force of 100 tons of TNT. This possibility led one economist to write, "Some plutonium in the hands of terrorists could make us long for the good old days when only 150 people at a time were held hostage in an airplane."

In 1976, John Phillips, a student at Princeton University in New Jersey, did several months of research, mostly in government publications, and designed a nuclear weapon. (He was not the first, or last college student to do so.) In the opinion of one nuclear physicist, Phillips had a workable plan for bomb-making. News about the successful design spread, and Phillips received a telephone call from a Pakistani official who asked for a copy! Phillips declined, but in reality such a public approach to bomb-making information is unnecessary. Plans for nuclear bombs are easily obtained. As one nuclear scientist put it, "The genie is out of the bottle."

Nuclear proponents believe that such scenarios are far-

In a film about plutonium made for television, actor Jack Lemmon demonstrates the basic "recipe" for a plutonium bomb.

In a demonstration to test the safety of nuclear fuel transport, a locomotive traveling over 80 miles per hour is deliberately crashed into a trailer carrying a cask used to ship nuclear fuel. The cask suffered only dents.

fetched, and just plain fear-mongering. The safeguards against sabotage and theft are so good, they believe, that only a large, well-organized, and heavily armed group could steal a nuclear shipment. And bomb-making is too difficult for a lone terrorist. Even a trained group would face tough problems, with good chances of killing only themselves. Besides, why go to so much trouble? Terrorists can more easily obtain poisons and other deadly weapons for their purposes.

Public concern about this danger led to stronger safeguards at existing nuclear plants. A 1974 study by the General Accounting Office found that "a security system at a licensed nuclear power plant could not prevent a takeover for sabotage by a small number—as few, perhaps, as two or three—of armed individuals."

Lax security was shown to exist at several power plants. In 1977, the NRC ordered better protection at reactors, enrichment plants, and other nuclear facilities. The steps included hiring more guards, training them better, and arming them with semiautomatic rifles.

On a global scale, a plutonium economy would make it possible for virtually any nation—or determined group within a nation—to develop nuclear weapons. According to a 1977 study by the NRC's Oak Ridge National Laboratory, a country with access to spent reactor fuel could build a simple, inexpensive reprocessing plant in six months, then separate enough plutonium from spent fuel to make a bomb in just a week. The laboratory report explained the process in detail; its release was criticized because it would help potential bomb-makers.

Nuclear weapons are great equalizers. A small nation with such weapons has destructive power formerly possessed only by large nations. Many countries are interested and in some cases eager to get nuclear arms for their security and prestige. India achieved this goal by diverting plutonium from a research reactor built with Canada's help.

It is debatable whether the withdrawal of the United States from breeder development will slow the spread of nuclear weapons. Some people believe this will actually reduce the nation's influence in nuclear matters. However, health and safety issues alone may be enough to delay and perhaps permanently block the use of plutonium and the LMFBR in the United States, unless the experience of other nations changes public attitudes about these issues.

The breeder prototypes in Europe and the Soviet Union have all had shutdowns because of damage to pipes and generators from the caustic sodium coolant. During its first two years of operation, France's Phenix breeder had a fine eighty-percent capacity record, but then it was shut down for several months because of leaks in the steam generators. France has continued to work on the Superphenix, a 1,200-megawatt breeder which was scheduled for completion about 1983. Britain planned to start construction of a 1,300-megawatt breeder.

In these nations there was optimistic talk about export-

The Phenix breeder reactor near Marcoule, France, has been operating since 1974.

ing breeders to other countries. But public opposition to breeders and to all nuclear energy is growing. And there were still many unresolved economic and technical questions. The initial success of the 250-megawatt Phenix breeder may not apply to the much bigger Superphenix, since the design of the new breeder is significantly different. So far prototype breeders appear to need thirty years of breeding in order to double their original fuel load. This doubling time will have to be reduced to about twelve years for breeders to be successful commercially.

People in the United States may witness the success or failure of fast-breeder technology from afar. However, even if the LMFBR is dead in the United States, the idea of *somehow* breeding nuclear fuel or at least increasing the usefulness of uranium is very much alive. In 1977, President Carter proposed an international study and search for alternatives to plutonium breeders.

The alternatives exist. Some were evaluated and rejected long ago by the AEC. One possible alternative is a nuclear technology based on thorium-232, a more abundant element than uranium. Thorium can be changed to the fissionable uranium-233. Another possibility, called a converter reactor or near-breeder, could stretch uranium reserves as much as five times. However, while more proliferation resistant, both of these alternatives still produce fuel for nuclear weapons.

A third possibility is called a homogeneous reactor. Its uranium or thorium fuel would be only fissionable in the reactor core. No fissionable material would leave the reactor. The homogeneous reactor would produce exactly as much fuel in its core as it needed to keep going. The proposed design of this reactor would be quite unlike those in use today. If it proved workable and desirable, virtually all of the world's present nuclear technology would be obsolete.

In the late 1970s, the United States was approaching an important turning point in its energy policies. Just a decade before, the AEC had pictured a future with a thousand reactors, including hundreds of fast breeders. The revised picture showed many fewer reactors and perhaps no fast breeders at all.

The first thirty years of nuclear development may have been a giant false step. But nonetheless, nuclear power may be part of our future. Besides the possibility of different kinds of breeders there is also the prospect of nuclear fusion.

In fusion, light atoms are forced together (fused) and give off energy that can then be used to generate electricity. Fusion may have some advantages over fission, but physicists admit that the kind of fusion process now under development is not the cleanest or most desirable. It depends on lithium at one stage of fuel production, and lithium is not abundant. Spent fuel would be a minor problem, compared with fission leftovers. But the fusion process would produce many neutrons, making the reactor itself intensely radioactive. Also, fusion is not expected

to be cheap. Once perfected, about the year 2000 at the earliest, fusion may prove to be more costly than solar energy or other alternatives now being investigated.

It seems likely that our future will include nuclear power. The energy is there in the atoms, humans are curious and inventive, and there are fortunes to be made in perfecting some kind of nuclear power. But it may bear almost no resemblance to the present troubled nuclear enterprise.

Physicists are investigating the feasibility of using powerful lasers to cause the nuclei of atoms to fuse briefly and release energy.

STOP NUKES
HOT DOG
CHILI DOG
COLD
ONLY
CONSUMER
NUCLEAR INDUSTRY
GOV'T
URANIUM
WILL
THIS IS
ENERGY
BREAK
WITH TRADITION
STOP
BEFORE A CATASTROPHE

5

NUCLEAR POLITICS

In 1978, there were 222 nuclear-power reactors in the world. In the United States, 71 reactors were operating and providing almost twelve percent of the nation's electricity. An impressive beginning.

Nuclear advocates were perplexed by the opposition they faced. They believed that nuclear energy was a necessity, and that the troublesome and "emotional" critics were only a temporary obstacle.

Temporary or not, the difficulties facing nuclear development were partly the result of important social changes in the United States and some other nations. People were concerned about the environment, especially as it affected their health. They were less trustful of institutions, especially of big business and government. They were skeptical of uncritical enthusiasm for a new, untested, and potentially damaging technology. They could point to DDT and a number of other modern "wonders" as examples of technological folly. They were wary of *suboptimization*

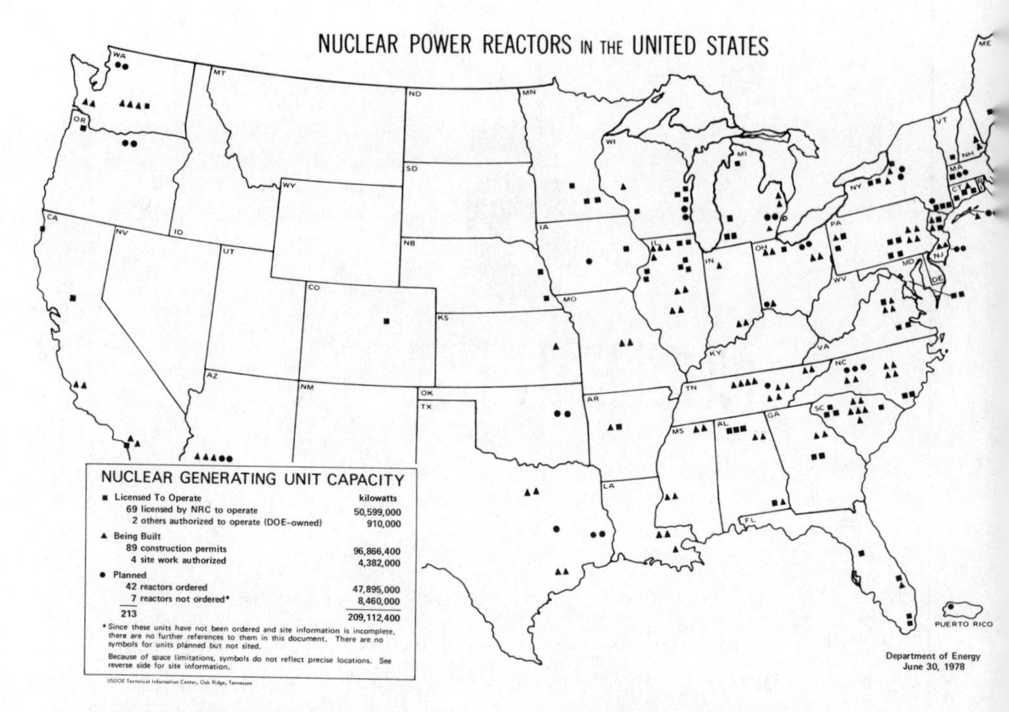

In 1978, seventy-one reactors produced electricity in the United States.

—finding the best possible way to do something which should not be done at all. And finally, many people were increasingly more willing to take political action, from picketing to lobbying, in order to achieve their goals.

This cultural change isn't about to evaporate overnight, and neither is the opposition to nuclear power. Antinuclear groups feel that time is on their side, and that government and industry efforts to "foist an unsafe technology on a trusting and uninformed public" will eventually fail. There is an increasing "erosion of ignorance" about nuclear power, and the more informed people are, the more opposition is expected to grow.

To some extent, the future of nuclear power may depend on changes in capital costs, on the outcome of safety and waste-storage tests, and on the development of alternate sources of energy. It also depends partly on luck. Both opponents and proponents agree that a major nuclear accident would probably cause such a public outcry that use of nuclear power would end abruptly.

Some people still argue that decisions about energy technology ought to be the exclusive province of engineers and scientists. But this was never really the case; energy choices have been economic and political choices. Besides, there is sharp disagreement among scientists about nuclear power. So, with increasing numbers of people feeling less inclined to "leave it to the experts," the future of nuclear power in the United States will be decided in courts, in Congress, in state legislatures, and in voting booths.

The year 1976 marked the emergence of a truly national antinuclear movement. Political action moved beyond the quiet realm of court suits and hearings about reactor licenses. There were peaceful demonstrations at construction sites and at actual nuclear plants. Citizens used the initiative process to get a sort of antinuclear proposal on election ballots in seven states (Arizona, California, Colorado, Montana, Ohio, Oregon, and Washington).

This process allows citizens to propose new laws directly. If a specific number of registered voters sign a petition to put a proposed law to a vote, the issue is then approved or disapproved in a statewide election. The wording of the proposed amendment varied from state to state, but each was aimed at having more proof of safety before nuclear development continued. The amendments were called "nuclear safeguard initiatives" by their supporters; "nuclear shutdown initiatives" by their detractors.

The proposals were voted down in all seven states, in most cases by a two-to-one margin. Spokesmen for the nuclear power industry hailed the results as a "green light" for nuclear development. But the 1976 initiatives were just round one of a long fight. (In the 1978 elections an antinuclear referendum was approved by voters in Montana.) Some observers believed that antinuclear forces had lost some battles, and would lose more, but were winning the war.

One environmentalist observed, "We're trying to overcome thirty years of industry and government propaganda. Even when these proposals go down to defeat, we've educated more millions of people about the problems we see and get more people on our side."

In the late 1970s, public opinion polls showed waning support for nuclear power. A Louis Harris poll in 1978 indicated that only forty-seven percent of the public favored speeding up nuclear development; a year earlier sixty-one percent had supported it. In a 1977 poll, just thirty-three percent of the people were unwilling to have a nuclear plant in their community. Soon after the Three Mile Island incident, fifty-six percent were opposed. Oppo-

You can tell by bumper stickers when you are near the construction site of a nuclear reactor.

sition to more nuclear plants built anywhere in the nation also increased.

Before and after the Three Mile Island incident, the emphasis of nuclear opponents was on safety issues. They succeeded in barring radioactive waste storage in several states. On the national level they sought a moratorium on further reactor licenses. But issues other than safety were involved and influenced the votes of the general public and of Congress.

A public-opinion specialist hired by Westinghouse said in 1976, "Just as the anti-nuclear, anti-growth, anti-technology forces have sold fear, the industry needs to find a lever with equal emotional intensity—massive unemployment, no growth, poorer living standards, runaway costs, and foreign dominance."

These were the issues emphasized in industry advertisements. For advertising during the 1976 round of state

Construction workers are not necessarily nuclear power advocates, but favor any large-scale project which provides jobs.

initiatives, the nuclear industry greatly outspent its opponents, who could not afford an equal advertising campaign. Some voters became convinced that nuclear energy was more of a pocketbook issue than a safety issue.

Rising oil prices and uncertainty about supplies were concerns of nuclear proponents. One United States senator said, "We must increase our use of nuclear power because we cannot afford to be dependent on foreign oil."

The link between oil and nuclear power is not this simple, however. Nuclear plants produce only electricity. It is no substitute for most uses of petroleum, such as heating oil and gasoline. Only one-eighth of United States oil consumption is used to generate electricity. So additional nuclear plants would not reduce oil imports much. The link between oil and electricity can be broken entirely by replacing oil-burning power plants with coal-burners.

In 1979, United States utilities had about thirty-two percent more generating capacity than was needed at times of peak demand. According to the President's Council on Environmental Quality, an effective energy conservation program would enable the nation to reach the year 2000 without building a single new power plant of any kind. This seems to be an unrealistic goal, but it illustrates that the future of nuclear power is tied in part to future demand for electricity. The fate of the nation's nuclear program may depend as much on economics (including costs of electricity to consumers) and on cultural attitudes towards saving energy as it does on safety issues.

One continuing issue of the nuclear controversy was the

Price-Anderson Act, which limited the nuclear industry's liability in the event of a catastrophe. Originally enacted in 1957, the act was renewed in 1967 and again, for ten more years, in 1975. This time, however, passage in Congress was not certain. Surprising numbers of congressmen and senators voted for amendments that would have allowed people to sue for damages above the specified liability limits. The amendments failed, but the act was changed somewhat to gradually phase out the government's obligations and to increase the liability of utilities. But the amount of liability would still fall far short of the potential cost of a major accident.

Although the Price-Anderson Act was extended to 1985, in 1977 it was declared unconstitutional by a federal court in North Carolina. Judge James McMillan reviewed the Rasmussen Report and declared that the mathematical odds for or against a nuclear catastrophe were not relevant. He wrote, "The significant conclusion is that under the odds quoted by either side, a nuclear catastrophe is a real, not fanciful, possibility."

He concluded that the act was unconstitutional in several ways, including that it "irrationally and unreasonably burdens victims of nuclear accidents to a greater extent than the law burdens victims of other accidents." Judge McMillan also concluded that the act tended to encourage irresponsibility in matters of safety, rather than encourage responsibility on the part of builders and owners.

The decision was quickly appealed to the Supreme Court, which ruled on the matter in June 1978. It reversed the lower-court decision and let the Price-Anderson Act stand. Chief Justice Warren E. Burger wrote that the liability limit is "an acceptable method for Congress to utilize in encouraging the private development of electrical energy by atomic power." Thus the government's goal of encouraging nuclear power "is ample justification" for placing nuclear accident victims in a special category.

In another case (involving construction of two nuclear plants), Justice William H. Rehnquist wrote, "Nuclear power may some day be a cheap, safe source of power, or it may not. But Congress has made a choice to at least try nuclear energy, establishing a reasonable review process in which courts are to play only a limited role."

Congress is the key. It is Congress (and the government agencies it created) that started, supported, and shielded the nuclear power industry, and helped hide its dangers from the public. So antinuclear individuals and groups tried even harder to effect change through Congress. Their efforts included attempts to greatly alter or do away with the Price-Anderson Act.

There has been lots of speculation about the impact of ending the Price-Anderson Act. Some people believed that it was vital to the nuclear-power business, and its end would be "one of the final nails in the nuclear industry's coffin." Others felt that the industry could go on without the protection of the act. Utilities would buy as much insurance as was available and trust in their own propaganda about the small risks of a serious accident.

If a major accident did occur, the entire financial bur-

"Your job might depend on a nuclear power plant like the one I help run." Roberta Kankus.

"The average American worker uses over 18 kilowatt hours of electricity every hour on the job. That's almost as much as the average family uses in one day!

"So, if you work for a living, you should care whether there'll be enough electricity available to keep your lathe or computer going. Or the lights in your office building.

"I believe the answer is to build more nuclear power plants. Nuclear power is safe. And it's cheaper and environmentally cleaner than most generating plants using other fuels.

"Very simply, what I want to get across is that nuclear power is a readily available source of energy the oil-producing countries can't shut off.

"I hope you'll read up on nuclear power. Understanding it can mean a lot to you—and your job."

Edison Electric Institute
for the electric companies
90 Park Avenue, New York, N.Y. 10016

Nuclear Engineer Roberta Kankus has received her Senior Operator License from the Nuclear Regulatory Commission to operate Philadelphia Electric's Peach Bottom Atomic Power Station.

den might not fall on utilities anyway, since an investigation might pin blame on a faulty piece of equipment. Manufacturers such as Westinghouse and General Electric have lobbied hardest for the protection provided by the Price-Anderson Act. Some observers, including a former chairman of the AEC, believed that an end to the Price-Anderson Act would have good effects on the quality of nuclear equipment. He said, "Do away with it and you'd probably see nuclear valves coming off the assembly line in a lot better shape."

If the Price-Anderson Act ends, nuclear equipment makers will probably seek another favorable law in Congress, perhaps one that would assign the full liability for the cost of a nuclear accident to taxpayers. There was a time when passage of such generous legislation was assured, but the once-solid pronuclear vote in Congress has been eroding year by year.

As the nuclear controversy moved further into the open political arena, both sides increased their efforts to influence public opinion. The nuclear industry tried, with partial success, to prevent showing of television programs which might provoke strong antinuclear reactions among viewers. It also sought advice from an opinion-research company, which concluded that women, blacks, young people, and the "less educated" were the "weakest

According to an opinion-research company, ads featuring young women scientists, like this one of the Edison Electric Institute, would help increase support for nuclear power.

link in the antinuclear coalition." These people could be "made to understand" by using advertisements featuring scientists who were women or young or black. The nuclear industry began using such advertisements, and also launched programs designed to influence students in colleges and high schools.

"Experts" were used by both sides of the controversy. A 1975 opinion poll had concluded that "for the final word on nuclear energy the public looks not to environmentalists, not to government leaders, and not to the media," but rather to "scientists—in fact, scientists inspired confidence in people on both sides of the fence." If the public seems confused about nuclear energy, it may be because people sense, correctly, that there is disagreement among the scientists.

One of the most intriguing and unexplored questions of the nuclear controversy is *why* a person chooses to be "for" or "against." (Neutrality seems to be a sign of lack of interest or information.)

Both sides of the nuclear controversy question the motives and sincerity of their opponents. People who urge a halt or a slowing down of nuclear development have been called irrational, illogical, hysterical, a tiny minority of nay-sayers, and a relatively small group of environmental no-growth muggers.

The Three Mile Island incident sparked many anti-nuclear demonstrations throughout the world.

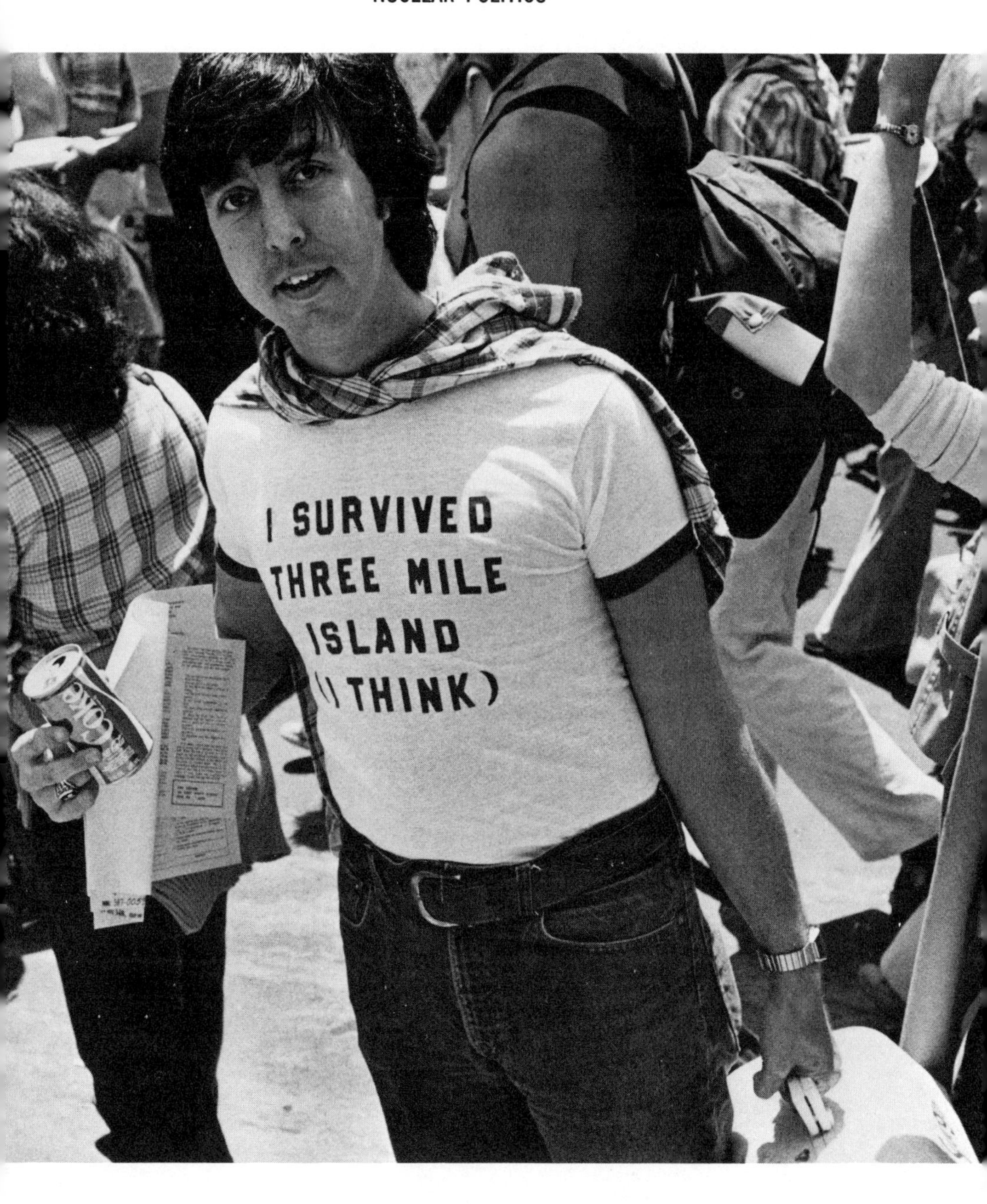
I SURVIVED
THREE MILE
ISLAND
(I THINK)

Antinuclear groups have a more exact explanation of their opponent's motives. They say that the most active support of nuclear development comes from people who have a direct or indirect stake in the nuclear enterprise, including careers, financial investments, and research grants. These people have strong motivations to make nuclear technology succeed. With careers and status at stake, it is predictable that nuclear engineers and the staffs of college nuclear-engineering departments would support nuclear energy. The same is true of staff members of nuclear industries and the government agencies that have a large commitment to nuclear power. (This makes all the more remarkable the resignation of some career engineers from within the nuclear business.)

Nuclear advocates tend to be optimistic about big technological ventures, and about the ability of science and technology to solve any problem, no matter how hazardous.

At times the motivations of some nuclear advocates are hidden from the public. In 1975, for example, a group of thirty-two scientists, mostly physicists, issued a "Scientists' Statement on Energy Policy." The statement urged the rapid expansion of nuclear power as the only realistic solution to energy needs. The statement was widely publicized by utilities and the nuclear-energy business.

Many people were probably impressed by the statement of these scientists. They included eleven Nobel Prize laureates, and fifteen other well-known professors from major universities. An investigation revealed, however, that most of this group were long-time supporters of nuclear power. Some of their research had been funded by the AEC. Fourteen of the twenty-six academic members of the group were affiliated with major corporations, including companies involved in the nuclear-energy business.

In 1975, there was a counterstatement by 2,300 scientists, engineers, and physicians who opposed nuclear development. Further announcements by scientists, for and against nuclear power, have followed. Each side has the support of some Nobel laureates. Each side looks for hidden motivations behind its opponents' attitudes, and interprets its own motives in a favorable way.

Consider, for example, the views of Dr. George Wald, a biology professor at Harvard and a Nobel laureate who opposes nuclear development. He concluded that the opinions of people who might lose money, privilege, or status by expressing their opinions were a little more credible than those of people who might gain money, privilege, or status.

Dr. Wald concluded, "Those of us who oppose nuclear power in its present forms have nothing to gain thereby but our share in the common good. Our opposition brings us into conflict with all the centers of power. It costs us our money. It threatens rather than raises our professional status."

This view suggests that the nuclear controversy is a clear-cut "good guys versus bad guys" conflict. But human behavior is not that simple. Both sides feel they are working "for the public good." But both sides may be motivated by personal interest or psychological factors, and both sides react emotionally.

Not much attention has been paid to the underlying motivations of either side and there is little solid information available. But it would be wise to keep these factors in mind and to regard claims of total neutrality and evenhandedness with suspicion. In the nuclear-power controversy, feelings influence the objectivity of everyone.

"Antinuclear" is a convenient label for individuals and groups, but it covers those who want to end the entire nuclear enterprise as well as those who urge a go-slow approach. For example, Dr. George Kistiakowsky, professor of chemistry at Harvard and a former presidential science advisor, believes that nuclear energy will eventually play a role in providing power. But, he said, "The technology is not ready for major acceleration. We are not ready. The world is not ready."

The distinction between a halt and a go-slow attitude seems lost to many nuclear advocates. As one scientist observed, "One must either buy the whole package—reprocessing, the breeder, and all—or else be considered an antagonist."

Some opponents of nuclear power feel empathy for the thousands of individuals whose jobs, status, and investments depend on nuclear energy. If the nuclear enterprise fails, as they hope, they believe that the federal government should ease the blow for individuals (and perhaps for corporations) who were attracted to the nuclear field, at least in part, by the government's pronuclear policies.

One of the main rationales for nuclear energy for the past thirty years has been its necessity. Many nuclear advocates sincerely believe that nuclear energy *must* be developed. As one nuclear engineer said, "Before long, nuclear power will be as essential to maintaining a safe and comfortable way of life as is oxygen in the air."

Nuclear advocates both in and out of government have tried to make prophecies like this come true. In advertisements and "information" booklets they express concern about future energy shortages while downplaying the potential of other power sources. More important, nuclear proponents have played a role in discouraging the development of alternative energy technologies. For decades, research and development funds have been devoted almost entirely to nuclear energy. In 1976, for example, ERDA's budget gave the breeder reactor nearly double the combined total amount of money allowed for research on energy conservation, and solar, wind, and geothermal energy.

Increasingly, however, the "necessity" of nuclear power in the United States is being questioned. In 1977, President Carter said, "The need for atomic power for peaceful purposes has been greatly exaggerated." In the 1979 federal budget, for the first time more than half of energy-research funds were devoted to nonnuclear sources and energy conservation. There was growing optimism about slowing energy demand, and speeding the development of solar power

ERDA
NASA

Opponents of nuclear power are encouraged by increased funding for the study of wind and solar energy.

and other alternatives. (Books and articles about alternatives to nuclear power are listed at the back of this book.)

The nuclear controversy will be bitterly contested for many years to come. People in the United States will hear many conflicting statements, claims, and statistics. They will be asked, through their representatives in Congress, to give the nuclear enterprise continued and perhaps increased financial help. They will also be asked to help slow down or halt nuclear development.

The people have an awesome responsibility. Their decisions will affect the abundance and cost of energy in the future. Their actions or inactions could also mean the difference between life and death for many thousands of people. They have the opportunity to learn as much as possible about nuclear energy, to make the tough decisions, and to settle the nuclear controversy themselves.

GLOSSARY

ALPHA PARTICLE—a high-energy particle, consisting of two neutrons and two protons, which is emitted by such radioactive atoms as plutonium and radon.

ATOM—the smallest unit of an element, consisting of a central nucleus, surrounded by orbiting electrons.

BETA PARTICLE—a high-energy electron emitted by the nucleus of a radioactive atom as it decays.

BREEDER—a nuclear reactor which produces energy and more fissionable nuclei than it consumes. See LIQUID METAL FAST BREEDER REACTOR.

CAPACITY FACTOR—the amount of power a generating plant actually produces, compared with the amount it ideally could produce. A capacity factor of eighty percent is quite good.

CHAIN REACTION—a series of nuclear fissions which are sustained because enough neutrons are produced to keep splitting nuclei and producing energy and more neutrons.

CONTROL ROD—a rod of neutron-absorbing material (such as boron or cadmium) inserted into a reactor's fuel core in order to soak up neutrons and shut off or slow down the rate of fission.

DECAY—the process of radioactive change in which atoms give off particles and change to atoms of lighter elements.

DECOMMISSIONING—the process of permanently shutting down and disposing of a nuclear power plant.

ELECTRON—a negatively charged particle of an atom which orbits its nucleus and is much lighter than a proton or neutron.

ENRICHMENT—the process of increasing to at least three percent the portion of uranium-235 in a quantity of uranium. The first enrichment method developed involved twelve hundred steps and was very expensive. A one-step enrichment process, using lasers, may be developed.

FALLOUT—radioactive fission products created by nuclear explosions, which fall from the atmosphere to the earth's surface.

FISSION—the process by which an atom nucleus splits and produces heat energy and radioactive particles. Nuclear fission produces the energy in nuclear power plants, and in some nuclear weapons. See also FUSION.

FISSION PRODUCTS—atoms formed as a result of the fissioning of other atoms. A variety of fission products form as uranium fissions in a reactor.

FUSION—the joining together of two light nuclei to form a single heavier nucleus and to produce energy. The sun's energy results from fusion of hydrogen nuclei.

GAMMA RAY—high-energy radiation of great penetrating power, emitted by nuclei of some radioactive elements.

HALF-LIFE—the period of time during which half the nuclei in a quantity of radioactive material undergo decay. Some fission products have half-lives measured in seconds; some in thousands of years.

ISOTOPE—an atom which has the same number of protons in its nucleus as other varieties of an element but has a different number of neutrons. The most common uranium isotope is uranium-238; the fissionable nuclear fuel is the isotope uranium-235.

LIQUID METAL FAST BREEDER REACTOR (LMFBR)—the breeder reactor that is currently the most developed in the world. It is cooled by liquid sodium, and "breeds" fuel as fast neutrons are absorbed by atoms of uranium-238 and changed to fissionable plutonium-239.

MEGAWATT—a thousand kilowatts, or a million watts, of electric power.

MELTDOWN—a possible accident in a reactor's core. As a result of overheating, part or all of the fuel would liquify and collapse. The most likely cause of a meltdown would be loss of the reactor's coolant.

NEUTRON—an uncharged particle of an atomic nucleus. Neutrons are ejected at high speed during fission and can be absorbed by another nucleus.

NUCLEUS—the positively charged center of an atom, made up of protons and neutrons. The great amounts of energy released when a nucleus is split is the source of nuclear power.

PLUTONIUM—a radioactive element not normally found in nature, plutonium is a reactor byproduct that can be used as reactor fuel. It emits alpha particles and is extremely dangerous when inhaled. Its half-life is 24,000 years.

PROTON—a positively charged particle which makes up part of the nucleus of an atom.

RADIOACTIVITY—behavior of a substance in which nuclei are undergoing change and emitting radiation in the form of alpha particles, beta particles, or gamma rays. This occurs naturally in a few elements and also can be produced artificially.

REM—a standard measurement of biological damage caused by radiation. It stands for "roentgen equivalent man."

REPROCESSING—treating spent reactor fuel in order to remove fission products and recover fissionable nuclei for use as fuel.

SUBOPTIMIZATION—as defined by economist Kenneth Boulding, "finding the best way to do something which should not be done at all."

TAILINGS—gray sand-like material left over from extraction of uranium from its ore. Tailings contain radium and emit radon gas, both of which give off alpha particles.

THERMAL POLLUTION—excess heat that may have harmful environmental effects when released into the air or water.

TRANSURANICS—elements which have heavier atoms than uranium, and very long half-lives. They include plutonium, neptunium, and americium.

URANIUM—a dark gray metal, once considered useless, which is fuel for pressurized-water nuclear reactors. Uranium emits alpha particles as radioactivity.

YELLOWCAKE—uranium oxide, the final product at uranium mills; the amount of uranium-235 in it must be increased before it is useful as reactor fuel.

FURTHER READING

This list of books and articles includes a wide range of opinions about nuclear matters. The title often reveals whether an author(s) has a pro- or antinuclear attitude. Be especially careful of any publication that claims to be without bias. To keep up-to-date on developments in the nuclear power controversy, I recommend these magazines: *Science* (published by the American Association for the Advancement of Science, 1515 Massachusetts Ave., N.W., Washington, D.C. 20005) and *The Bulletin of the Atomic Scientists* (1020-24 East 58th St., Chicago, Illinois 60637). Both publish articles and news reports which represent views of all sides of the nuclear debate; this is not true of *Scientific American*, which through early 1979 had published mostly articles of nuclear advocacy by long-time proponents of nuclear power. In all periodicals watch for follow-up letters commenting on published articles.

ALEXANDER, TOM, "Why the Breeder Reactor Is Inevitable." *Fortune,* September 1977, pp. 123-130.

AMERICAN NUCLEAR SOCIETY, *Nuclear Power and the Environment,* Second Edition. LaGrange Park, Illinois: 1976.

ANGINO, ERNEST, "High-Level and Long-Lived Radioactive Waste Disposal." *Science,* December 2, 1977, pp. 885-890.

BEBBINGTON, WILLIAM, "The Reprocessing of Nuclear Fuels." *Scientific American,* December 1976, pp. 30-41.

BERGER, JOHN, *Nuclear Power: The Unviable Option.* Palo Alto, California: Ramparts Press, 1976.

BETHE, HANS, "The Necessity of Fission Power." *Scientific American,* January 1976, pp. 21-31.

BOFFEY, PHILIP, "Nuclear Power Debate: Signing Up the Pros and Cons." *Science,* April 9, 1976, pp. 120-122.

BUPP, IRVIN, *et al.,* "The Economics of Nuclear Power." *Technology Review,* February 1975, pp. 15-21.

BUPP, IRVIN, and DERAIN, JEAN-CLAUDE, *Light Water: How the Nuclear Dream Dissolved.* New York: Basic Books, 1978.

CARTER, LUTHER, "Radioactive Wastes: Some Urgent Unfinished Business." *Science,* February 18, 1977, pp. 661-666 and 704.

CARTER, LUTHER, "Political Fallout from Three Mile Island." *Science,* April 13, 1979, pp. 154-155.

CARTER, LUTHER, "Nuclear Wastes: The Science of Geological Disposal Seen as Weak." *Science,* June 9, 1978, pp. 1135-1137.

CASPER, BARRY, "Laser Enrichment: A New Path to Proliferation?" *Bulletin of the Atomic Scientists,* January 1977, pp. 28-41.

FURTHER READING

CENTER FOR SCIENCE IN THE PUBLIC INTEREST, *People and Energy*. Washington, D.C.: 1976.

CENTER FOR THE STUDY OF RESPONSIVE LAW, *A Citizen's Manual on Nuclear Energy*. Washington, D.C.: 1974.

CHOW, BRIAN, "The Economic Issues of the Fast Breeder Reactor Program." *Science*, February 11, 1977, pp. 551-556.

CLARK, WILSON, *Energy for Survival*. Garden City, New York: Anchor Press, 1975.

COCHRAN, THOMAS B., *The Liquid Metal Fast Breeder Reactor: An Environmental and Economic Critique*. Baltimore: Johns Hopkins University Press, 1974.

COHEN, BERNARD, "The Disposal of Radioactive Wastes from Fission Reactors." *Scientific American*, June 1977, pp. 21-31.

COMMONER, BARRY, *The Politics of Energy*. New York: Alfred A. Knopf, Inc., 1979.

COMMONER, BARRY, *The Poverty of Power*. New York: Alfred A. Knopf, Inc., 1976.

EBBIN, STEVEN, and KASPER, RAPHAEL, *Citizen Groups in the Nuclear Power Controversy*. Cambridge, Mass.: MIT Press, 1974.

FEIVESON, HAROLD, *et al.*, "The Plutonium Economy: Why We Should Wait and Why We Can Wait." *Bulletin of the Atomic Scientists*, December 1976, pp. 10-14.

FORD FOUNDATION, Energy Policy Project, *A Time To Choose: America's Energy Future*. Cambridge, Mass.: Ballinger Publishing Company, 1974.

GOFMAN, JOHN, and TAMPLIN, ARTHUR, *Poisoned Power: The Case Against Nuclear Power Plants*. Emmaus, Pa.: Rodale Press, Inc., 1971.

GWYNNE, PETER, "Plutonium: 'Free' Fuel or Invitation to Catastrophe?" *Smithsonian,* July 1976, pp. 93-98.

HARWOOD, STEVEN, *et al.,* "The Cost of Turning It Off." *Environment,* December 1976, pp. 17-20, 25-26.

HAYES, DENIS, *Energy: The Case for Conservation,* Worldwatch Paper 4. Washington, D.C.: Worldwatch Institute, 1976.

HAYES, DENIS, *Rays of Hope: The Transition to a Post-Petroleum World.* New York: W. W. Norton & Co., Inc., 1977.

HOHENEMSER, CHRISTOPH, *et al.,* "The Distrust of Nuclear Power." *Science,* April 1, 1977, pp. 25-34.

KERR, RICHARD, "Geologic Disposal of Nuclear Wastes: Salt's Lead Is Challenged." *Science,* May 11, 1979, pp. 603-606.

KERR, RICHARD, "Nuclear Waste Disposal: Alternatives to Solidification in Glass Proposed." *Science,* April 20, 1979, pp. 289-291.

KNELMAN, FRED, *Nuclear Power: The Unforgiving Technology.* Edmonton, Alberta, Canada: Hurtig Publishers, 1976.

MARSHALL, ELIOT, "Assessing the Damage at TMI." *Science,* May 11, 1979, pp. 594-596.

MARX, JEAN, "Low-Level Radiation: Just How Bad Is It?" *Science,* April 13, 1979, pp. 160-164.

MCCAULL, JULIAN, "The Cost of Nuclear Power." *Environment,* December 1976, pp. 11-16.

MCCLARY, STEVEN, and PRIMACK, JOEL, "Breeder Reactors, the Biggest Nuclear Gamble." *Sierra Club Bulletin,* March 1977, pp. 12-16.

MCPHEE, JOHN, *The Curve of Binding Energy.* New York: Farrar, Straus & Giroux, 1974.

METZ, WILLIAM, "Reprocessing Alternatives: The Options Multiply." *Science*, April 15, 1977, pp. 284-287.

METZ, WILLIAM, "Report of Fusion Breakthrough Proves to Be a Media Event." *Science*, September 1, 1978, pp. 792-794.

METZGER, H. PETER, *The Atomic Establishment*. New York: Simon & Schuster, 1972.

MURPHY, ARTHUR, ed., *The Nuclear Power Controversy*. Englewood Cliffs, New Jersey: Prentice-Hall, Inc., 1976.

NADER, RALPH, and ABBOTT, JOHN, *The Menace of Atomic Energy*. New York: W. W. Norton & Company, 1977.

NOVICK, SHELDON, *The Careless Atom*. Boston: Houghton Mifflin, 1969.

NYE, JOSEPH, "Time to Plan for the Next Generation of Nuclear Technology." *Bulletin of the Atomic Scientists*, October 1977, pp. 38-41.

OLSON, MCKINLEY, *Unacceptable Risk: The Nuclear Power Controversy*. New York: Bantam Books, 1976.

PATTERSON, WALTER, *Nuclear Power*. New York: Penguin Books, 1976.

POST, R.F., and RIBE, F.L., "Fusion Reactors as Future Energy Sources." *Science*, November 1, 1974, pp. 397-407.

PRINGLE, LAURENCE, *Energy: Power for People*. New York: Macmillan Publishing Co., Inc., 1975.

ROCHLIN, GENE, "Nuclear Waste Disposal: Two Social Criteria." *Science*, January 7, 1977, pp. 23-31.

———, *et al.*, "West Valley: Remnant of the AEC." *Bulletin of the Atomic Scientists*, January 1978, pp. 17-26.

SCIENTIST'S INSTITUTE FOR PUBLIC INFORMATION, *Nuclear Power: Economics and the Environment.* Oakland, California: 1976.

SHAPLEY, DEBORAH, "Reactor Safety: Independence of Rasmussen Study Doubted." *Science,* July 1, 1977, pp. 29-31.

———, "Nuclear Power Plants: Why Do Some Work Better Than Others?" *Science,* March 25, 1977, pp. 1311-1313.

U.S. NUCLEAR REGULATORY COMMISSION, *Reactor Safety Study* (Rasmussen Report). Washington, D.C.: 1975.

VENDRYES, GEORGES, "Superphenix: A Full-Scale Breeder Reactor." *Scientific American,* March 1977, pp. 26-35.

VON HIPPEL, FRANK, "Looking Back on the Rasmussen Report." *Bulletin of the Atomic Scientists,* February 1977, pp. 42-47.

———, and WILLIAMS, ROBERT, "Toward a Solar Civilization." *Bulletin of the Atomic Scientists,* October 1977, pp. 12-15, 56-60.

WEBB, RICHARD, *The Accident Hazards of Nuclear Power Plants.* Amherst, Mass.: The University of Massachusetts Press, 1976.

WEINBERG, ALVIN, "Is Nuclear Energy Acceptable?" *Bulletin of the Atomic Scientists,* April 1977, pp. 54-60.

INDEX

Asterisk (*) indicates photograph or drawing

advertisements about nuclear power, 18, 105, 107, *110, 112, 117
alpha particles, 22, 26
American Physical Society, 37
americium, 61, 62
Atomic Energy Commission (AEC): actions of, 9–12, 26, 29, 30, 33, 34–36, 38, 39, 44, 49, 65–66, 72, 77, 78, 84, 88, 96, 97, 107, 111, 115; creation of, 9; end of, 12, 39
Atomic Industrial Forum, 34
atoms, 14, *15

Babcock & Wilcox Company, 50
beta particles, 22
boron, 16
breeder reactors, 26, 34, 36, 44, 62, 82, 83–84, *85, 86, *87, 88–89, 94, *95, 96–97, 116
Britain, 64, 86, 89, 94
Brookhaven National Laboratory, 11, 33
Brookhaven Report, 11, 33
Brooks, Harvey, 59
Browns Ferry incident, 41–43
Burger, Warren E., 109

cadmium, 16
Canada, 64, 70–71, 82, 94
cancer, 11, 24, 26, 30, 53, 62
capacity factor, 77–78, 80
capital costs, 75–77, 87, 103
Carter, President, 47, 85–86, 96, 117
cesium, 61
chain reaction, *15, 16, 86
China syndrome, 32
Clinch River breeder, 84, *85
coal, 6, 15, 16, 22, 53, 55, 58, 61, 76, 78, *79, 80, 81, 86, 107
Cohen, Bernard, 68–69, 75
Congress (U.S.), 9, 12, 34, 35, 47, 54, 65, 77, 103, 105, 108, 109, 111, 120
conservation of energy, 6, 7, 107, 117
control rods, 16–17, 32, 41, 44
cooling towers, *56, 57
Council on Economic Priorities, 78, 80
Council on Environmental Quality, 67, 107

decommissioning of old reactors, 71–73
Department of Energy, 39, 66
Department of Labor, 53

electricity: production of, 16, *17, 47, 50, 55, 75, *87, 97, 101, 107; use of, 6, 76, 77, 107
Emergency Core–Cooling Systems (ECCS), 18, 38–39, 41
Energy Research and Development Administration (ERDA), 39, 66–67, 80, 117
Enrico Fermi Breeder Reactor, 44, 84, *88
Environmental Protection Agency, 30, 38

fission, 14–*15, 16, 21, 26, 32, 33, 47, 55, 58, 69, 73, 84
fission products, 15, 21, 32, 33, 61, 65
fossil fuels, 6, 22, 55, 58, 76, 89; *see also* coal, oil
France, 64, 82, 86, 89, 94
Freedom of Information Act, 34
fuel rods, 16, 17, *20, *28, 46, 59, *60
fusion, 97–99

gamma rays, 22
General Accounting Office (GAO), 49, 66–67, 93
General Electric, 12, 45, 111
Germany (West), 64, 71

half-lives of radioactive elements, 22, *23, 33, 61, 62, 65, 73
Hendrie, Joseph M., 47, 49
Hydrogen gas bubble, 46, 47

initiative process, 103–104, 107
insurance against nuclear accident, 10–12; *see also* Price-Anderson Act
iodine, 15, 59

Japan, 7, 64, 71, 86

Kistiakowsky, George, 116
Komanoff, Charles, 78, 80

Lemmon, Jack, *91
Liquid Metal Fast Breeder Reactor (LMFBR), 84–*87, 88–89, 94, *95, 96–97
lithium, 97
Loss of Coolant Accident (LOCA), 32–33, 38, 87

Mattson Roger J., Dr., 49
McMillan, James, 108
meltdown of reactor core, 32–33, 36, 38, 41, 44, 46, 47, 88
motivation of people for or against nuclear power, 112, 114–116

National Academy of Sciences, 30, 70
National Committee on Radiation Protection, 30
National Engineering Laboratory, 35, 39, *40
neptunium, 61
neutrons, 15, 16, 17, 33, 84, 86, 97
nickel, 73
Nixon, Richard, 84–85
Nuclear Age, 7, 12
nuclear wastes, 54–55, 58–73, 105
nuclear power: and politics, 9, 12, 34–35, 39, 66, 71, 73, 94, 101–120; as weapons, 7, *8, 9, 10, 63, 82, 85, 86, 90, *91, 93–94, 96; controversy over, 1–2, 6–7, 27, 32, 101–120; cost of, 10, 62, 71, 75–77, 80–82, 85; fuel cycle of, 7, 18, *21, 30, 36, 53–55, 62, 66, 73, 89; fuel supply for, 81–84, 86, 96; safety of, 11, 18, 29–51, 53, 58, 69, 72–73, 77, 87–90, *92, 93–94, 103, 107
Nuclear Regulatory Commission (NRC), 30, 38, 39, 44, 45, 47, 49, 50, 57, 73, 77, 93, 107

Oak Ridge National Laboratory, 93–94
oil, 5, 6, 16, 22, 86, 107

Petroleum Age, 5
Phenix breeder, 87, *95
Phillips, John, 90
plutonium, 21, 22, 24, *25, 26, 27, 33, 55, 61, 62, 63, 65, 82, 86, 89, 90
Pollard, Robert, 44–45, 49
Price-Anderson Act, 12, 33, 108–109, 111; *see also* insurance

Public Health Service, 53
public opinion, 10, 14, 49, 51, 59, 63, 73, 101–103, 104–105, 111

radioactivity: and cancer, 11, 24, 30, 32, 53, 62; defined, 22; in fallout, 10; in and near power plants, 17, 29–30, 46–47, 50, 71–73, 78; natural, 24, 69
radon, 53, 54
Rasmussen, Norman, 35–37
Rasmussen Report, 36–38, 108
Ray, Dixie Lee, 59
reactors, 16, *17, 18, 21, *28, 30, 33, 34, 35, 36, 38, 39, 41, 43, 47, 50, 61, 73, 93, 96–97; *see also* breeder reactors
Reactor Safety Study, *see* Rasmussen Report
Rehnquist, William H., 109
rem, 23, 24, 30
reprocessing of fuel, 21–22, 58–59, 62–65, 82, 84, 89, 93–94
Resnikoff, Martin, 73
Rockefeller, Nelson, 58

salt formations for waste storage, 65, 66, 67, *68–71
San Joaquin Nuclear Project, 57–58
scientists' opinions, 30, 33, 34, 37, 38, 69, 70, 75, 90, 103, 112, 114–115
Sierra Club, 37
sodium (as coolant), 84, *87, 88, 89, 94
solar energy, 6, 99, 117, *119
Soviet Union, 9, 10, 44, 86, 94
storage of wastes, 21, 55, 59, 61, 62, 65, 71, 105
strontium, 15, 22–23, 61
Superphenix, 94, 96

tailings, 54, 69, 89
thermal pollution, 55, 57, 58
thorium, 54, 96
Thornburg, Richard, 46, 47, 51
Three Mile Island incident, 46–51, 77, 81, 104, 105
Three Mile Island nuclear plant, *viii
transuranics, 61–62, 65

Union of Concerned Scientists, 37, 39, 107
uranium, 15, 16, 21, 22, 24, 26, 32, 33, 62, 81–84, 86, 89, 96
uranium hexafluoride, 18, 54–55
uranium mines, 18, *52, 53, 69, 82, 89
uranium oxide pellets, 16, *19, 21
utilities, 10, 11, 12, 18, 36, 38, 46, 55, 57, 76, 77, 78, 80, 81, 107, 109, 111

Vermont Yankee Nuclear Plant, 47

Wald, George, 115
wastes, *see* nuclear wastes
water, cooling, 16, 17, 29, 32, 55, 57, 58
Westinghouse Electric Corporation, 12, 81–82, 105, 111

yellowcake, 18

PICTURE CREDITS:

Edison Electric Institute, 110; Engelhardt in the *St. Louis Post-Dispatch*, 72; French Embassy Press & Information Division, 95; Herblock Cartoons in *The Washington Post*, 64; Laurence Pringle, 4, 74, 105; Sean Pringle, 100, 113. Sacramento Municipal Utility District, 56; Tennessee Vailey Authority, 19 (bottom), 31, 42; Don Widener (KCET-TV), 91; Wide World Photos, viii, 106. Drawings and diagrams by Publishing Synthesis, Ltd. (adapted from U.S. Department of Energy).

The following were supplied by the U.S. Department of Energy, courtesy of: Battelle Northwest Laboratory, 70; E. I. DuPont de Nemours & Company, 60; ERDA's Oak Ridge National Laboratory, 68; FEA Photo, 79; Iowa Electric Light and Power Company, 28; Lawrence Livermore Laboratory, 98; Lookout Mountain Air Force Station, 8; NASA Photo, 118; Phillips Petroleum Company, 40; Raymond Reczek, 63; Rochester Gas & Electric Company, 48; Rockwell International, 25; U.S. Department of Energy, 52, 83, 85, 88, 92, 102 (map), 119; Westinghouse Electric Corporation, 13, (Atomic Power Division) 19 (top), (Jack Merhaut) 20.